Frequency Domain Hybrid Finite Element Methods for Electromagnetics

Frequency Domain Hybrid Finite Element Methods for Electromagnetics
John L. Volakis, Kubilay Sertel and Brian C. Usner

ISBN: 978-3-031-00566-4   paperback
ISBN: 978-3-031-00566-4   paperback

ISBN: 978-3-031-01694-3   ebook
ISBN: 978-3-031-01694-3   ebook

DOI 10.1007/978-3-031-01694-3

A Publication in the Springer series
SYNTHESIS LECTURES ON COMPUTATIONAL ELECTROMAGNETICS

Series ISSN: 1932-1252   print
Series ISSN: 1932-1716   electronic

Series Editor: Constantine A. Balanis, Arizona State University

First Edition
10 9 8 7 6 5 4 3 2 1

Frequency Domain Hybrid Finite Element Methods for Electromagnetics

John L. Volakis, and Kubilay Sertel
The Ohio State University, Columbus, Ohio, USA

Brian C. Usner
Ansoft Corporation

SYNTHESIS LECTURES ON COMPUTATIONAL ELECTROMAGNETICS

ABSTRACT

This book provides a brief overview of the popular Finite Element Method (FEM) and its hybrid versions for electromagnetics with applications to radar scattering, antennas and arrays, guided structures, microwave components, frequency selective surfaces, periodic media, and RF materials characterizations and related topics. It starts by presenting concepts based on Hilbert and Sobolev spaces as well as *Curl* and *Divergence* spaces for generating matrices, useful in all engineering simulation methods. It then proceeds to present applications of the finite element and finite element-boundary integral methods for scattering and radiation. Applications to periodic media, metamaterials and bandgap structures are also included. The hybrid volume integral equation method for high contrast dielectrics and is presented for the first time. Another unique feature of the book is the inclusion of design optimization techniques and their integration within commercial numerical analysis packages for shape and material design. To aid the reader with the method's utility, an entire chapter is devoted to two-dimensional problems. The book can be considered as an update on the latest developments since the publication of our earlier book (*Finite Element Method for Electromagnetics*, IEEE Press, 1998). The latter is certainly complementary companion to this one.

KEYWORDS

Finite Elements, Integral Equations, Volume Integral Methods, Hybrid Techniques, Numerical Methods, Antennas, Radiation, Radar Scattering, Electromagnetics, Periodic media, Metamaterials, Design, Optimization

Contents

1. **Introduction** 1
   1.1 Essentials of Computational Methods in EM 1
       1.1.1 Concepts from Functional Analysis 2
       1.1.2 Variational Techniques for the Solution of PDEs and IEs 5
       1.1.3 $\mathcal{H}(\mathrm{Curl}; \Omega)$ and $\mathcal{H}(\mathrm{Div}; \Omega)$ Spaces 7
       1.1.4 Maxwell's Equations 10
   1.2 Parametric Geometry Fundamentals 12
       1.2.1 Parametric Geometry in 1D 13
       1.2.2 Parametric Geometry in 2D 15
       1.2.3 Parametric Geometry in 3D 18
   1.3 Curvilinear Finite Elements 19
       1.3.1 Divergence Conforming Elements 20
       1.3.2 Curl Conforming Elements 21
   1.4 Overview 24

2. **Two-Dimensional Hybrid FE–BI** 25
   2.1 The Boundary Value Problem 26
       2.1.1 TE Polarization 27
       2.1.2 TM Polarization 28
       2.1.3 Boundary Conditions 28
   2.2 Surface Equivalence and Boundary Integral Equations 33
   2.3 Variational Formulation 34
   2.4 Discretization 36
   2.5 Example Discretization 39
   2.6 2D Scattering Applications 42

3. **Three-Dimensional Hybrid FE–BI: Formulation and Applications** 51
   3.1 The Boundary Value Problem 52
   3.2 Boundary Integral Equations 53
   3.3 The FE–BI Variational Statement 54
   3.4 Discretization 55

3.5   Applications 58
    3.5.1   Scattering 58
    3.5.2   Antenna Radiation 62

4.   Hybrid Volume-Surface Integral Equation 69
  4.1   Generalized VSIE Formulation 70
  4.2   Boundary Conditions 76
  4.3   Variational Form of the VSIE 77
  4.4   Discretization 78
    4.4.1   Junction Resolution 80
    4.4.2   MoM System Development 82
  4.5   Examples 84
    4.5.1   Junction Resolution Validation 84
    4.5.2   Scattering Examples 85
    4.5.3   Antenna Examples 90

5.   Periodic Structures 93
  5.1   Periodic Boundary Conditions 94
    5.1.1   Using the FE–BI 96
    5.1.2   Using the VSIE 98

6.   Antenna Design and Optimization Using FE–BI Methods 109
  6.1   Design Optimization: Overview 111
    6.1.1   Definition 111
    6.1.2   Classification 111
  6.2   Design Examples 118
    6.2.1   Example 1: Dielectric Material Optimization of a Patch Antenna
        via Topology Optimization and SLP 118
    6.2.2   Example 2: Optimization of an Irregular-shaped Dual-band
        Patch Antenna via SA and GA 122
    6.2.3   Multiobjective Antenna Design Using Volumetric Material
        Optimization and Genetic Algorithms 124
  6.3   Comments 129

Preface

This book was started with the goal of providing a brief overview of the popular finite element method (FEM) and its hybrid versions for electromagnetics with applications to radar scattering, antennas and arrays, guided structures, microwave components, frequency selective surfaces, periodic media, and RF materials characterizations to mention a few. However, as the project evolved we realized that several developments had occurred since the publication of the book *Finite Element Method for Electromagnetics: Antennas, Microwave Circuits, and Scattering Applications (1998)* coauthored by Volakis, Chatterjee and Kempel. Thus, we enhanced this book to also include an update to applications and methods that occurred over the past few years. More specifically, it also includes applications of finite element–boundary integral (FE–BI) methods to infinite and finite periodic media and introduces the hybrid volume integral equations for high contrast dielectrics and metamaterials. The importance of design optimization and its integration within commercial numerical analysis packages is recognized by the inclusion of a chapter written by a former Ph.D. student Professor Gullu Kiziltas and current postdoctoral researcher Dr. Stavros Koulouridis.

Given the routine use of numerical methods for large-scale modeling, the book starts with theoretical concepts for generating robust matrix systems from partial differential equations (PDEs) and integral equations (IEs). Hilbert and Sobolev spaces as well as $\mathcal{H}(\text{Curl}; \Omega)$ and $\mathcal{H}(\text{Div}; \Omega)$ spaces are presented in simple engineering terms and various basis functions, including parametric, curvilinear, curl and divergence conforming as well as higher order are concisely presented in the context of the Galerkin and Petrov–Galerkin methods for casting PDEs and IEs into discrete systems. This is followed by Chapter 2, which gives a step-by-step development of the FEM and its hybrid FE–BI version for two-dimensional scattering applications with considerations for resistive and conductive (or magnetic) cards as well as impedance boundary conditions. Several examples are given with enough details for the reader to repeat them. Chapter 3 gives a three-dimensional overview of the FE–BI method with applications to scattering, conformal antennas and large finite arrays where the repeatability of the unit cell is exploited to reduce memory and computational resources.

Chapters 4 and 5 are devoted to the hybrid volume integral equations, including the volume surface integral equations (VSIEs). These integral equations are particularly suited for high contrast dielectrics and can also be combined with PDE-based matrix systems to yield the most efficient and robust hybrid method. Chapter 4 gives the mathematical details (including matrix element calculations) with simple applications for verification and others for displaying their greater capability. Chapter 5 is particularly devoted to periodic media, including frequency selective surfaces (FSS) and frequency selective volumes (FSV) as well as metamaterial and electromagnetic bandgap structures. Examples of periodic media that yield negative constitutive parameters are discussed and their bandgap features examined. Finally, Chapter 6 starts by giving a brief survey of gradient (namely, sequential quadratic or linear programming) and stochastic (namely, genetic algorithms) optimization methods used in RF design. The second half of Chapter 6 is devoted to shape, topology, and material design optimization examples for antennas using the hybrid FE–BI methods as the solver within the optimization algorithm.

John L. Volakis
Kubilay Sertel
Brian C. Usner
July 2, 2006
The Ohio State University
Columbus, Ohio, USA

CHAPTER 1

Introduction

A number of numerical solutions for electromagnetics appeared in the 1960s employing moment methods [1,2] and finite element techniques [3–5]. However, the application of these early implementations was very limited. With the availability of mainframe computing machines in the 1970s and 1980s, focus was primarily on high-frequency techniques [6] and related approximations that could handle practical problems using available CPU resources. Significant focus was seen in the 1980s on all methods, including the moment method [7] and finite difference methods [8, 9]. In regard to the finite element method, the focus was on developing approximate absorber boundary conditions for truncating the finite element mesh as close as possible to the scatterer or radiator [10, 11]. Concurrently, research on robust iterative solvers grew substantially in the 1980s and continued vigorously in the 1990s. Hybrid methods, involving a combination of integral, finite element, and possibly high-frequency methods, have always been considered [12, 13] as a means of minimizing computational requirements by employing each method where it is most efficient.

1.1 ESSENTIALS OF COMPUTATIONAL METHODS IN EM

We begin by first discussing the mathematical architecture germain to any numerical method for electromagnetic analysis. The idea is to familiarize the reader with the general framework upon which most finite element and integral equation implementations are built. In recent years, researchers have constructed an elegant mathematical foundation that enables accurate discretization of Maxwell's equations. We refer the reader to [14, 15] and the references therein for a review of this work. However, we do note that most of this work was pioneered even earlier by Nedelec [16] (who first presented divergence and curl conforming finite elements) and Whitney [17, 18]. Below, we begin by introducing the reader to the components essential for developing any numerical method. This discussion is not meant to be complete from a mathematical standpoint, but rather to introduce the reader to concepts and literature on the subject. The application of these mathematical frameworks into practical engineering tools took several more years to develop, and to a great degree they followed developments in computing capabilities including memory and CPU speed.

1.1.1   Concepts from Functional Analysis

We will soon see that numerical solutions of electromagnetic equations—be it partial differential equations (PDEs), integral equations (IEs), or some hybrid thereof—arise from a consideration of their variational or weak form. When one uses the variational form to solve an equation, it is first necessary to show the solution's existence and uniqueness. Additionally, it is essential to derive some type of error bound between the exact solution and the numerically computed solution. Without such proof or guarantee, any endeavor in hopes of finding a solution is likely to be futile. To answer such questions, one must first develop a mathematical framework based upon which existence and uniqueness theorems can be derived, and this is done below.

Hilbert Spaces

To begin with, we restrict our discussion to a special vector function space called a Hilbert space. We do this because functions contained within Hilbert spaces are most easily applicable to real-world computational schemes. Though there are more general function spaces (e.g., Banach spaces) that satisfy similar properties as the Hilbert space, their consideration is not necessary in our context.

To define the Hilbert space, let us consider vector function space $\mathcal{V}$ over complex numbers $\mathbb{C}$ having domain $\Omega \subset \mathbb{R}^d$, where $\mathbb{R}^d$ is the d-dimensional space of real numbers $\mathbb{R}$. If the scalar product (often called an inner product) on $\mathcal{V}$ is defined to be a map $\langle \cdot, \cdot \rangle : \mathcal{V} \times \mathcal{V} \longrightarrow \mathbb{C}$ such that

1. if $\mathbf{v} \in \mathcal{V}$ then $\langle \mathbf{v}, \mathbf{v} \rangle = 0$ if and only if $\mathbf{v} \equiv 0$,

2. $\langle \mathbf{u}, \mathbf{v} \rangle = \langle \mathbf{v}, \mathbf{u} \rangle^*$ for all $\mathbf{u}, \mathbf{v} \in \mathcal{V}$, where * denotes a complex conjugate,

3. $\langle c_1 \mathbf{u}_1 + c_2 \mathbf{u}_2, \mathbf{v} \rangle = c_1 \langle \mathbf{u}_1, \mathbf{v} \rangle + c_2 \langle \mathbf{u}_2, \mathbf{v} \rangle$ for all $\mathbf{u}_1, \mathbf{u}_2, \mathbf{v} \in \mathcal{V}$ and $c_1, c_2 \in \mathbb{C}$,

we then say that $\mathcal{V}$ is a Hilbert space provided it is complete with respect to the norm $||\mathbf{v}|| = \sqrt{\langle \mathbf{v}, \mathbf{v} \rangle}$. Here, completeness refers to whether or not a Cauchy sequence [19] converges when the norm is used as the distance measure. That is, if the sequence of vector functions $\{\mathbf{v}_1, \mathbf{v}_2, \ldots\}$ is a Cauchy sequence, then completeness implies that $\lim_{n \to \infty} ||\mathbf{v}_n - \mathbf{v}_{n-1}|| = 0$. Above, we have only defined the properties of the scalar product, though it should be of no surprise that we are mostly interested in the Lebesgue scalar product, or $\mathcal{L}^2(\Omega)$ scalar product, given by

$$\langle \mathbf{u}, \mathbf{v} \rangle = \int_\Omega \mathbf{u}^* \mathbf{v} \ d\Omega, \qquad (1.1)$$

where * implies the adjoint operator (complex conjugate transpose).

We define an operator $A : \mathcal{V} \longrightarrow \mathcal{W}$, where $\mathcal{V}$ and $\mathcal{W}$ are Hilbert spaces, to be *linear* if

$$A(c_1 \mathbf{v}_1 + c_2 \mathbf{v}_2) = c_1 A \mathbf{v}_1 + c_2 A \mathbf{v}_2, \quad \forall c_1, c_2 \in \mathbb{C}, \quad \mathbf{v}_1, \mathbf{v}_2 \in \mathcal{V}. \qquad (1.2)$$

Furthermore, we say A is *bounded* if there exists a constant C such that $||A\mathbf{v}|| \leq C||\mathbf{v}||$ for all $\mathbf{v} \in \mathcal{V}$. We remark that a linear operator is bounded if and only if the linear operator is *continuous* (i.e., the topological structure of $\mathcal{V}$ is preserved in $\mathcal{W}$).

Dual space and linear functionals. For any Hilbert space $\mathcal{V}$, we can define the *dual space $\mathcal{V}'$* to be the space of bounded linear functionals on $\mathcal{V}$. If $\mathcal{V}$ represents the space of n-dimensional column vectors, then $\mathcal{V}'$ would be the space of n-dimensional row vectors because a row vector applied to a column vector results in a scalar value. For the Hilbert space defined above, these linear functionals are closely related to the scalar product. In other words, the linear functional $f \in \mathcal{V}'$ acting on $\mathbf{v} \in \mathcal{V}$ can be written as

$$f(\mathbf{v}) = \langle \mathbf{f}, \mathbf{v} \rangle, \tag{1.3}$$

for some vector function $\mathbf{f}$. Consequently, in the context of this discussion, we can equivalently say that $\mathcal{V}'$ is made up of all such vector functions $\mathbf{f}$ such that (1.3) is bounded for all $\mathbf{v} \in \mathcal{V}$.

Interestingly, when $\dim(\mathcal{V}) = n$ is finite and the set $\{\mathbf{v}_1, \mathbf{v}_2, \ldots, \mathbf{v}_n\}$ forms a basis for $\mathcal{V}$, then the $\dim(\mathcal{V}') = n$ and the corresponding basis for $\mathcal{V}'$ is given by $\{\mathbf{v}_1', \mathbf{v}_2', \ldots, \mathbf{v}_n'\}$ where

$$\langle \mathbf{v}_i', \mathbf{v}_j \rangle = \delta_{ij} \tag{1.4}$$

and δ_{ij} is the Kronecker delta-function. This important result is exploited when we consider the solution of the second-kind integral equations over curvilinear finite elements.

Sesquilinear forms and coercivity. Moving right along, let $\mathcal{V}$ and $\mathcal{W}$ be two Hilbert spaces, then the mapping $a(\cdot, \cdot) : \mathcal{V} \times \mathcal{W} \longrightarrow \mathbb{C}$ is called a *sesquilinear form* if

1. $a(\mathbf{v}_1 + \mathbf{v}_2, \mathbf{w}_1 + \mathbf{w}_2) = a(\mathbf{v}_1, \mathbf{w}_1) + a(\mathbf{v}_1, \mathbf{w}_2) + a(\mathbf{v}_2, \mathbf{w}_1) + a(\mathbf{v}_2, \mathbf{w}_2)$,

2. $a(c\ \mathbf{v}, \mathbf{w}) = c^* a(\mathbf{v}, \mathbf{w})$,

3. $a(\mathbf{v}, c\ \mathbf{w}) = c\ a(\mathbf{v}, \mathbf{w})$,

for all $\mathbf{v} \in \mathcal{V}$, $\mathbf{w} \in \mathcal{W}$, and $c \in \mathbb{C}$. Note that if $a(\cdot, \cdot)$ is a mapping into the space $\mathbb{R}$ instead of $\mathbb{C}$, then we call $a(\cdot, \cdot)$ a *bilinear form*. Similar to linear operators, we say that a sesquilinear form $a(\cdot, \cdot)$ is bounded if there exists a constant C such that $|a(\mathbf{v}, \mathbf{w})| \leq C||\mathbf{v}||\ ||\mathbf{w}||$ for all $\mathbf{v} \in \mathcal{V}$ and $\mathbf{w} \in \mathcal{W}$. Additionally, $a(\cdot, \cdot)$ defined on $\mathcal{V} \times \mathcal{V}$ is said to be *coercive* if a constant $C > 0$ exists such that $|a(\mathbf{v}, \mathbf{v})| \geq C||\mathbf{v}||^2$ for all $\mathbf{v} \in \mathcal{V}$. The coercivity property of sesquilinear forms is related to the positive definite properties of linear/matrix operators. This is easily seen when we consider the bilinear form $b(\cdot, \cdot) : \mathcal{V} \times \mathcal{V} \longrightarrow \mathbb{R}$ to be

$$b(\mathbf{u}, \mathbf{v}) = \langle \mathbf{u}, A\mathbf{v} \rangle = \int_{\Omega} \mathbf{u}^* A\mathbf{v}\ d\Omega. \tag{1.5}$$

We can readily observe from (1.5) that if A is positive definite, then $b(\cdot, \cdot)$ must be coercive.

Finite-dimensional subspaces: existence and uniqueness of numerical solutions.   We are now ready to state one of the most important results in functional analysis pertaining to the numerical solutions of Maxwell's equations. The following theorem, called Cea's theorem [20], is the finite-dimensional analog of the Lax–Milgram theorem, and the reader is referred to [14] for its proof.

Theorem 1.   *Suppose the space $\mathcal{V}_h \subset \mathcal{V}$, $h > 0$, is a family of finite-dimensional subspaces of a Hilbert space $\mathcal{V}$ parameterized by h, which in finite element implementations is directly related to the maximum diameter of the elements in a finite element mesh. Suppose $a : \mathcal{V} \times \mathcal{V} \longrightarrow \mathbb{C}$ is a bounded, coercive sesquilinear form and $\mathbf{f} \in \mathcal{V}'$. Then the problem of finding $\mathbf{v}_h \in \mathcal{V}_h$ such that*

$$a(\mathbf{t}_h, \mathbf{v}_h) = \langle \mathbf{t}_h, \mathbf{f} \rangle, \qquad \text{for all} \quad \mathbf{t}_h \in \mathcal{V}_h \qquad (1.6)$$

has a unique solution. Additionally, if $\mathbf{v} \in \mathcal{V}$ is the exact solution solving

$$a(\mathbf{t}, \mathbf{v}) = \langle \mathbf{t}, \mathbf{f} \rangle, \qquad \text{for all} \quad \mathbf{t} \in \mathcal{V} \qquad (1.7)$$

then there is a constant $C \in \mathbb{R}$ such that

$$||\mathbf{v} - \mathbf{v}_h|| \leq C \inf_{\mathbf{u}_h \in \mathcal{V}_h} ||\mathbf{v} - \mathbf{u}_h||, \qquad (1.8)$$

where inf is the "infimum" or greatest lower bound [19].

Essentially, the above theorem ensures that a solution to the discretized infinite space problem in (1.7) exists, is unique, and is bounded. It should be noted that as $h \to 0$ we expect $\dim(\mathcal{V}_h) \to \infty$. Additionally, we say that the space $\mathcal{V}_h$ is *conforming* to $\mathcal{V}$ if $\mathcal{V}_h \subset \mathcal{V}$ for every h.

Two key assumptions on the property of the sesquilinear form were made in Theorem 1, namely that the form be bounded and coercive. We will discuss the boundedness property in Section 1.1.3 where we formulate special types of Hilbert spaces upon which typical electromagnetic operators find a home (i.e., the curl and divergence conforming Sobolev spaces). Here, however, we mention that for the solution of Maxwell's equations it is not always possible to ensure the coercivity property of a sesquilinear form, or for that matter the positive definiteness of the linear operator defining the sesquilinear form. This is especially the case for the hyperbolic wave equation considered in finite element solutions and for the second-kind integral equations (magnetic field and volume integral equations) found in scattering applications. Thus, an additional theorem is necessary to prove existence, uniqueness, and boundedness for these cases. To do so, it is necessary to consider the so-called Fredholm alternative theorem [14].

Theorem 2.   *Let $B : \mathcal{V} \longrightarrow \mathcal{V}$ be a bounded linear operator, where B is a compact operator and I is the identity operator. Then either*

1. *the homogeneous equation* $(I + B)\mathbf{v} = 0$ *has exactly p linearly independent solutions for some finite integer $p > 0$, or*

2. *for every* $\mathbf{f} \in \mathcal{V}$, *the inhomogeneous equation* $(I + B)\mathbf{v} = \mathbf{f}$ *has a unique solution.*

Once an operator can be described in the form $I + B$, from Theorem 2, existence and uniqueness follow once we prove that -1 is not an eigenvalue of B.

Because the second-kind integral equation operators are by definition of form $I + B$ in Theorem 2, we next present one last theorem adapted from [21]. This theorem is similar to Theorem 1 since it describes the conditions upon which one can guarantee the existence, uniqueness, and boundedness of the discretized problem of finding a $\mathbf{v} \in \mathcal{V}$ such that

$$(I + B)\mathbf{v} = \mathbf{f}, \tag{1.9}$$

where $\mathbf{f} \in \mathcal{V}$ and $B : \mathcal{V} \longrightarrow \mathcal{V}$ is a compact linear operator not having -1 as an eigenvalue.

Theorem 3. *Suppose $\mathcal{V}_h \subset \mathcal{V}$, $h > 0$, is a family of conformal finite-dimensional subspaces of a Hilbert space $\mathcal{V}$. Further, suppose $B : \mathcal{V} \longrightarrow \mathcal{V}$ is a compact linear operator not having -1 as an eigenvalue. Then the problem of finding $\mathbf{v}_h \in \mathcal{V}_h$ such that*

$$\langle \mathbf{t}_h, (I + B)\mathbf{v}_h \rangle = \langle \mathbf{t}_h, \mathbf{f} \rangle, \qquad \text{for all} \quad \mathbf{t}_h \in \mathcal{T}_h \subset \mathcal{V}' \tag{1.10}$$

has a unique solution if and only if $\dim(\mathcal{V}_h) = \dim(\mathcal{T}_h)$ and $\mathcal{T}_h \bigcap \mathcal{V}_h^\perp = \{\emptyset\}$. Additionally, if $\mathbf{v} \in \mathcal{V}$ is the exact solution of (1.9) then there is a constant C such that

$$||\mathbf{v} - \mathbf{v}_h|| \leq C \inf_{\mathbf{u}_h \in \mathcal{V}_h} ||\mathbf{v} - \mathbf{u}_h|| \tag{1.11}$$

for sufficiently small h.

One of the most important differences between Theorems 1 and 3 relates to the space which the testing functions $\mathbf{t}_h$ are chosen from. Specifically, for the solution of equations that can be represented in the form given in (1.9), the testing functions should be chosen to lie in a space dual to that of the solution space. This should be contrasted to Theorem 1 where the testing functions are selected from the solution space itself. This result has profound effects when choosing testing functions for the second-kind integral equations, such as the volume integral equation (VIE) over curvilinear finite elements and for the construction of symmetric hybrid formulations [22].

1.1.2 Variational Techniques for the Solution of PDEs and IEs

In this section, we show how the above theorems can be used in casting Maxwell's equations into a numerical set of equations. We first exploit Theorem 1 to explain Galerkin's method for solving variational problems. We then employ Theorem 3 to explain the Petrov–Galerkin method for solving variational problems arising from the second-kind integral equations.

Galerkin's Method

We begin by considering the linear operator $A : \mathcal{V} \longrightarrow \mathcal{V}'$ and the inhomogeneous system

$$A\mathbf{v} = \mathbf{f}, \tag{1.12}$$

where $\mathbf{v} \in \mathcal{V}$ and $\mathbf{f} \in \mathcal{V}'$. One can think of operator A as the PDE operator corresponding to the vector wave equation used, for example, in finite element solutions of Maxwell's equations or the first-kind integral equation operator, called the electric field integral equation (EFIE), used in computing the radar scattering for conducting targets. The source term $\mathbf{f}$ can be thought to represent the incident or excitation field for radar scattering or the feed model for antenna radiation problems. Because the construction of the inverse operator A^{-1} is not particularly obvious, to solve (1.12) we must first consider its variational form. That is, we seek $\mathbf{v} \in \mathcal{V}$ such that

$$\langle \mathbf{t}, A\mathbf{v} \rangle = \langle \mathbf{t}, \mathbf{f} \rangle, \qquad \forall \mathbf{t} \in \mathcal{V}. \tag{1.13}$$

This variational form enforces (1.12) with respect to the testing functions $\mathbf{t}$, and is often referred to as its weak form. We note that the term on the left-hand side of (1.13) is a sesquilinear form similar to that defined in (1.5). We also assume that A is bounded and positive definite, or is such that one can prove the existence and uniqueness of the variational form in (1.13). Let us also assume that the n-dimensional space $\mathcal{V}_h \subset \mathcal{V}$ is described by the basis $\{\mathbf{v}_1, \mathbf{v}_2, \ldots, \mathbf{v}_n\}$, such that $\mathbf{v}_h = \sum_{i=1}^{n} \alpha_i \mathbf{v}_i$. If we seek $\mathbf{v}_h \in \mathcal{V}_h$ such that

$$\sum_{i=1}^{n} \alpha_i \langle \mathbf{t}_h, A\mathbf{v}_i \rangle = \langle \mathbf{t}_h, \mathbf{f} \rangle, \qquad \forall \mathbf{t}_h \in \mathcal{V}_h, \tag{1.14}$$

then from Theorem 1, or in some cases with the help of Theorem 2 [14], $\mathbf{v}_h$ exists, and is unique, and is bounded. To ensure that (1.14) holds for all $\mathbf{t}_h \in \mathcal{V}_h$, it only suffices to select n linearly independent vector testing functions (testing basis) from $\mathcal{V}_h$. If these vector bases are chosen to be the same as the basis used to describe $\mathbf{v}_h$, the resulting numerical scheme is then called *Galerkin's method*. That is, we seek $\mathbf{v}_h \in \mathcal{V}_h$ such that

$$\sum_{i=1}^{n} \alpha_i \langle \mathbf{v}_j, A\mathbf{v}_i \rangle = \langle \mathbf{v}_j, \mathbf{f} \rangle, \qquad \forall \mathbf{v}_j, \quad j = 1, 2, \ldots, n. \tag{1.15}$$

This system (1.12) can next be turned into an n-dimensional linear set of equations whose solution is found by solving the $(n \times n)$-matrix system

$$\bar{A}\mathbf{x} = \mathbf{b}. \tag{1.16}$$

Here, the matrix entries are given by $\bar{A}_{ji} = \langle \mathbf{v}_j, A\mathbf{v}_i \rangle$, the source vector is $\mathbf{b}_j = \langle \mathbf{v}_j, \mathbf{f} \rangle$, and the unknown $\mathbf{x}$ contains the unknown expansion coefficients α_i in (1.15).

Petrov–Galerkin Method

We now consider the numerical solution of the system described in (1.9), where we reiterate that the operator $B : \mathcal{V} \longrightarrow \mathcal{V}$ and the excitation term $\mathbf{f} \in \mathcal{V}$. Again, because the inverse operator $(I + B)^{-1}$ is not particularly obvious, we must consider its variational or weak form in which we seek $\mathbf{v} \in \mathcal{V}$ such that

$$\langle \mathbf{t}, (I + B)\mathbf{v} \rangle = \langle \mathbf{t}, \mathbf{f} \rangle, \qquad \forall \mathbf{t} \in \mathcal{V}'. \tag{1.17}$$

As was done in the preceding section, we consider the n-dimensional space $\mathcal{V}_h \subset \mathcal{V}$ having basis $\{\mathbf{v}_1, \mathbf{v}_2, \ldots, \mathbf{v}_n\}$, such that $\mathbf{v}_h = \sum_{i=1}^{n} \alpha_i \mathbf{v}_i$. If we seek the numerical solution $\mathbf{v}_h \in \mathcal{V}_h$ such that

$$\sum_{i=1}^{n} \alpha_i \langle \mathbf{t}_h, (I + B)\mathbf{v}_i \rangle = \langle \mathbf{t}_h, \mathbf{f} \rangle, \qquad \forall \mathbf{t}_h \in \mathcal{T}_h \subset \mathcal{V}', \tag{1.18}$$

then as long as $\dim(\mathcal{T}_h) = n$ and $\mathcal{T}_h \bigcap \mathcal{V}_h^{\perp} = \{\emptyset\}$ Theorem 3 ensures that the solution $\mathbf{v}_h$ is unique and bounded for sufficiently small h. One should note that conditions on the testing space $\mathcal{T}_h$ are much less restrictive than those in the preceding section where we considered Galerkin's testing. This fact allows for all sorts of testing schemes from point collocation [21] to interesting meshless methods found in the current literature [23]. In our work [24], we have used (1.4) to construct a proper set of testing functions $\{\mathbf{t}_1, \mathbf{t}_2, \ldots, \mathbf{t}_n\}$ for the second-kind volume integral equation (VIE) over curvilinear finite elements. As such, the *Petrov–Galerkin* statement consists finding a $\mathbf{v}_h \in \mathcal{V}_h$ such that

$$\sum_{i=1}^{n} \alpha_i \langle \mathbf{t}_j, (I + B)\mathbf{v}_i \rangle = \langle \mathbf{t}_j, \mathbf{f} \rangle, \qquad \forall \mathbf{t}_j, \quad j = 1, 2, \ldots, n, \tag{1.19}$$

to be recast into the $(n \times n)$-linear system

$$\bar{M}\mathbf{x} = \mathbf{b}, \tag{1.20}$$

where the matrix entries are given by $\bar{M}_{ji} = \langle \mathbf{t}_j, \mathbf{v}_i \rangle + \langle \mathbf{t}_j, B\mathbf{v}_i \rangle$, the source vector is given by $\mathbf{b}_j = \langle \mathbf{t}_j, \mathbf{f} \rangle$, and the unknown vector $\mathbf{x}$ contains the expansion coefficients α_i in (1.19).

1.1.3 $\mathcal{H}(\text{Curl}; \Omega)$ and $\mathcal{H}(\text{Div}; \Omega)$ Spaces

We saw in Section 1.1.1 that to ensure unique and bounded numerical solutions, one of the main assumptions was that the linear operator be bounded (i.e., continuous) over its domain space. Thus, it seems rather important to define the meaning of "bounded." Subsequently, we will describe the special Hilbert spaces, called Sobolev spaces, over which one can ensure that the differential operators (e.g., the divergence and curl operators) found in Maxwell's equations are also bounded.

Let us consider k times continuously differentiable scalar functions over domain Ω, denoted by $C^k(\Omega)$, and the corresponding D-dimensional vector function space $(C^k(\Omega))^D$, then for up to certain values k we can ensure continuity for any differential operator. This is a statement often encountered in a differential equation course. However, when the differential equation is cast in a numerical system this space becomes much too restrictive because the finite element subspaces are often piecewise continuous at best. Thus, we need a less restrictive manner to describe the meaning of "boundedness." However, we caution the reader not to assume that the definition of boundedness is arbitrary. In fact, the definition is intimately linked to the scalar product, and therefore the norm, defined in the Hilbert space.

The $\mathcal{L}^2(\Omega)$ Lebesgue Space

The space of square integrable functions, called the Lebesgue space, defines boundedness according to the $\mathcal{L}^2(\Omega)$ scalar product defined in (1.1). Specifically, we define the D-dimensional Lebesgue space $(\mathcal{L}^2(\Omega))^D$ to be a Hilbert space such that

$$(\mathcal{L}^2(\Omega))^D = \{\mathbf{v} : \Omega \to \mathbb{C}^D \mid ||\mathbf{v}||_{(\mathcal{L}^2(\Omega))^D} = \int_\Omega |\mathbf{v}|^2 d\Omega < \infty\}. \qquad (1.21)$$

It can be seen from this definition that $(\mathcal{L}^2(\Omega))^D$ contains sets of functions, namely piecewise continuous functions, that can be used within finite element implementations. However, to guarantee boundedness of a differential operator over the domain of square integrable functions we must further restrict ourselves to spaces associated with derivatives which are also square integrable. These vector spaces are referred to as Sobolev spaces, and the reader is referred to [25] for a complete discussion of their importance in the solution of differential equations.

The $\mathcal{H}^1(\Omega)$ Sobolev Space

One of the Sobolev spaces of importance to this text contains all scalar functions having a square integrable gradient. We denote this D-dimensional space as $\mathcal{H}^1(\Omega)$, mathematically defined as

$$\mathcal{H}^1(\Omega) = \{\phi \in \mathcal{L}^2(\Omega) \mid \nabla\phi \in (\mathcal{L}^2(\Omega))^D\} \qquad (1.22)$$

with the norm

$$||\phi||_{\mathcal{H}^1(\Omega)} = (||\phi||^2_{\mathcal{L}^2(\Omega)} + ||\nabla\phi||^2_{(\mathcal{L}^2(\Omega))^D})^{1/2}. \qquad (1.23)$$

This space is "home," if you will, to the space of functions that represent the scalar potentials used to represent the electric and magnetic fields in Maxwell's equations.

The $\mathcal{H}(\text{Curl}; \Omega)$ Sobolev Space

Another important Sobolev space to consider is that containing functions with a square integrable curl. In this context, we define $\mathcal{H}(\text{Curl}; \Omega)$ to be

$$\mathcal{H}(\text{Curl}; \Omega) = \{\mathbf{v} \in (\mathcal{L}^2(\Omega))^D \mid \nabla \times \mathbf{v} \in (\mathcal{L}^2(\Omega))^D\} \qquad (1.24)$$

with the norm

$$||\mathbf{v}||_{\mathcal{H}(\text{Curl};\Omega)} = (||\mathbf{v}||^2_{(\mathcal{L}^2(\Omega))^D} + ||\nabla \times \mathbf{v}||^2_{(\mathcal{L}^2(\Omega))^D})^{1/2}. \qquad (1.25)$$

As will be seen, this space plays an important role in describing the electric and magnetic field intensities (**E** and **H**) within Maxwell's equations.

The $\mathcal{H}(\text{Div}; \Omega)$ Sobolev Space

The Sobolev space containing those functions with square integrable divergence is defined as

$$\mathcal{H}(\text{Div}; \Omega) = \{\mathbf{v} \in (\mathcal{L}^2(\Omega))^D \mid \nabla \cdot \mathbf{v} \in \mathcal{L}^2(\Omega)\} \qquad (1.26)$$

with the norm

$$||\mathbf{v}||_{\mathcal{H}(\text{Div};\Omega)} = (||\mathbf{v}||^2_{(\mathcal{L}^2(\Omega))^D} + ||\nabla \cdot \mathbf{v}||^2_{\mathcal{L}^2(\Omega)})^{1/2}. \qquad (1.27)$$

This space is "home" to flux quantities such as the electric and magnetic flux densities (**D** and **B**) and current quantities (**M** and **J**).

de Rham Diagram

We present in this section an interesting result pertaining to the Sobolev spaces defined above. The well-known identities $\nabla \times \nabla \phi = 0$ and $\nabla \cdot \nabla \times \mathbf{v} = 0$, where $\phi \in \mathcal{H}^1(\Omega)$ and $\mathbf{v} \in \mathcal{H}(\text{Curl}; \Omega)$, serve to establish a relationship between the Sobolov spaces defined above. Specifically, we note that in view of the identity $\nabla \times \nabla \phi = 0$, $\nabla \phi \in \mathcal{H}(\text{Curl}; \Omega)$. Likewise, the identity $\nabla \cdot \nabla \times \mathbf{v} = 0$ implies that $\nabla \times \mathbf{v} \in \mathcal{H}(\text{Div}; \Omega)$. These relations lead to the well-known diagram given by

$$\mathcal{H}^1(\Omega) \xrightarrow{\nabla} \mathcal{H}(\text{Curl}; \Omega) \xrightarrow{\nabla \times} \mathcal{H}(\text{Div}; \Omega) \xrightarrow{\nabla \cdot} \mathcal{L}^2(\Omega). \qquad (1.28)$$

This diagram displays that the gradient operator maps functions from $\mathcal{H}^1(\Omega)$ to $\mathcal{H}(\text{Curl}; \Omega)$, whereas the curl operator maps functions from $\mathcal{H}(\text{Curl}; \Omega)$ into $\mathcal{H}(\text{Div}; \Omega)$.

Operators on Bounded $\mathcal{H}(\text{Curl}; \Omega)$ and $\mathcal{H}(\text{Div}; \Omega)$ Spaces

Due to practical computational constraints, most scattering and radiation problems are forced to consider finite computational domains Ω. Here the outer boundary of Ω will be denoted as $\partial\Omega$ and we constrain ourselves to only those $\partial\Omega$ that meet the Lipschitz condition [26]. Essentially, this condition implies that almost everywhere (a.e.) on the boundary $\partial\Omega$ there is a unique, well-defined unit normal $\hat{n}$ [14], or that the boundary can be represented everywhere by the graph of a function [27].

As can be expected, we are very much interested in the behavior of electromagnetic quantities on $\partial\Omega$. To aid us in the description of this behavior, we define two operators on $\partial\Omega$, namely the perpendicular surface trace operator $\gamma_\times$ and the tangential surface trace operator γ_t. Here, the perpendicular surface trace operator applied to a vector function $\mathbf{v} : \Omega \to \mathbb{C}^D$ is defined as

$$\gamma_\times \mathbf{v} = \hat{n} \times \mathbf{v}|_{\partial\Omega}. \tag{1.29}$$

If $\mathbf{v} \in \mathcal{H}(\mathrm{Curl}; \Omega)$, then $\gamma_\times : \mathcal{H}(\mathrm{Curl}; \Omega) \to \mathcal{H}^{-1/2}(\mathrm{Div}; \partial\Omega)$ (note that superscript$^{-1/2}$ is a notation denoting that the function is evaluated on a boundary and is adopted from the trace theorem [14, 22]). That is, we define the vector space $\mathcal{H}^{-1/2}(\mathrm{Div}; \partial\Omega)$ to be

$$\mathcal{H}^{-1/2}(\mathrm{Div}; \partial\Omega) = \{\mathbf{f} : \partial\Omega \to \mathbb{C}^D \mid \text{ there exists } \mathbf{v} \in \mathcal{H}(\mathrm{Curl}; \Omega) \text{ with } \gamma_\times \mathbf{v} = \mathbf{f}\}, \tag{1.30}$$

or more constructively

$$\mathcal{H}^{-1/2}(\mathrm{Div}; \partial\Omega) = \{\mathbf{f} \in (\mathcal{L}^2(\partial\Omega))^D \mid \hat{n} \cdot \mathbf{f} = 0 \text{ almost everywhere on } \partial\Omega, \nabla_{\partial\Omega} \cdot \mathbf{f} \in \mathcal{L}^2(\partial\Omega)\}. \tag{1.31}$$

Additionally, the tangential surface trace operator represents the boundary value of a vector function $\mathbf{v} : \Omega \to \mathbb{C}^D$, and is defined as

$$\gamma_t \mathbf{v} = \hat{n} \times (\mathbf{v}|_{\partial\Omega} \times \hat{n}). \tag{1.32}$$

Again, if $\mathbf{v} \in \mathcal{H}(\mathrm{Curl}; \Omega)$, then $\gamma_t : \mathcal{H}(\mathrm{Curl}; \Omega) \to \mathcal{H}^{-1/2}(\mathrm{Curl}; \partial\Omega)$. That is, we define the vector space $\mathcal{H}^{-1/2}(\mathrm{Curl}; \partial\Omega)$ to be

$$\mathcal{H}^{-1/2}(\mathrm{Curl}; \partial\Omega) = \{\mathbf{f} : \partial\Omega \to \mathbb{C}^D \mid \text{ there exists } \mathbf{v} \in \mathcal{H}(\mathrm{Curl}; \Omega) \text{ with } \gamma_t \mathbf{v} = \mathbf{f}\}, \tag{1.33}$$

or more explicitly

$$\mathcal{H}^{-1/2}(\mathrm{Curl}; \partial\Omega) = \{\mathbf{f} \in (\mathcal{L}^2(\partial\Omega))^D \mid \hat{n} \cdot \mathbf{f} = 0 \text{ almost everywhere on } \partial\Omega,$$
$$\nabla_{\partial\Omega} \times \mathbf{f} \in (\mathcal{L}^2(\partial\Omega))^D\}. \tag{1.34}$$

An important relation exists between the two spaces $\mathcal{H}^{-1/2}(\mathrm{Div}; \partial\Omega)$ and $\mathcal{H}^{-1/2}(\mathrm{Curl}; \partial\Omega)$. Specifically, it is possible to show that the two spaces are dual to one another, i.e., $(\mathcal{H}^{-1/2}(\mathrm{Curl}; \partial\Omega))' = \mathcal{H}^{-1/2}(\mathrm{Div}; \partial\Omega)$ [14]. In closing, we state below some useful properties of the two operators defined in (1.29) and (1.32)

$$\gamma_t \gamma_t = \gamma_t \tag{1.35}$$
$$\gamma_t \gamma_\times = \gamma_\times \tag{1.36}$$
$$\gamma_\times \gamma_t = \gamma_\times \tag{1.37}$$
$$\gamma_\times \gamma_\times = -\gamma_t. \tag{1.38}$$

1.1.4   Maxwell's Equations

Having defined the Sobolev spaces above, we next specialize them to the time harmonic electromagnetic field quantities. Assuming an $e^{j\omega t}$ time dependence, Maxwell's equations are succinctly

TABLE 1.1: Symbolism of the Electromagnetic Quantities and Their Units

SYMBOL	QUANTITY	UNITS	SYMBOL	QUANTITY	UNITS
E	Electric field intensity	V/m	**H**	Magnetic field intensity	A/m
D	Electric flux density	C/m^2	**B**	Magnetic flux density	T
J	Electric current density	A/m^2	**M**	Magnetic current density	V/m^2
ρ_e	Electric charge density	C/m^3	ρ_m	Magnetic charge density	T/m

stated as [28, 29]

$$\nabla \times \mathbf{E} = -j\omega\mathbf{B} - \mathbf{M}, \tag{1.39}$$
$$\nabla \times \mathbf{H} = j\omega\mathbf{D} + \mathbf{J}, \tag{1.40}$$
$$\nabla \cdot \mathbf{D} = \rho_e, \tag{1.41}$$
$$\nabla \cdot \mathbf{B} = \rho_m, \tag{1.42}$$

with the electromagnetic quantities and their units as stated in Table 1.1. The flux densities and field intensities are related through their respective constitutive relations given by

$$\mathbf{D} = \varepsilon_0\bar{\bar{\varepsilon}}_r \cdot \mathbf{E}, \tag{1.43}$$
$$\mathbf{B} = \mu_0\bar{\bar{\mu}}_r \cdot \mathbf{H}, \tag{1.44}$$

in which ε_0 and μ_0 are the intrinsic permittivity and permeability, respectively, whereas $\bar{\bar{\varepsilon}}_r$ and $\bar{\bar{\mu}}_r$ represent the 3×3 tensors for the relative permittivity and permeability of an arbitrary anisotropic material.

The function spaces which allow for proper discretization of the electromagnetic quantities are in Table 1.2. Starting with the electric and magnetic field intensities, $\mathbf{E}$ and $\mathbf{H}$, we see from (1.39) and (1.40) that the presence of the curl operator requires $\mathbf{E} \in \mathcal{H}(\text{Curl}; \Omega)$ and $\mathbf{H} \in \mathcal{H}(\text{Curl}; \Omega)$. Invoking the de Rham diagram in (1.28), namely that the curl operator maps functions from $\mathcal{H}(\text{Curl}; \Omega)$ to $\mathcal{H}(\text{Div}; \Omega)$, we then see that $\mathbf{D}, \mathbf{B} \in \mathcal{H}(\text{Div}; \Omega)$ and $\mathbf{J}, \mathbf{M} \in \mathcal{H}(\text{Div}; \Omega)$. In a similar way, the de Rham diagram can be used in conjunction with (1.41) and (1.42) to require that $\rho_e, \rho_m \in \mathcal{L}^2(\Omega)$. Additionally, the constitutive relations given in (1.43) and (1.44) imply that the dyadics $\varepsilon_0\bar{\bar{\varepsilon}}_r$ and $\mu_0\bar{\bar{\mu}}_r$ represent mappings from $\mathcal{H}(\text{Curl}; \Omega)$ to $\mathcal{H}(\text{Div}; \Omega)$. These observations are summarized in Table 1.2.

Before closing, we point out an important relationship between the spaces $\mathcal{H}(\text{Curl}; \Omega)$ and $\mathcal{H}(\text{Div}; \Omega)$. Particularly, if we consider the energy functionals in electromagnetics [29] (namely, the power stored in the electric field $\mathbf{E} \cdot \mathbf{D}$, the power stored in the magnetic field $\mathbf{H} \cdot \mathbf{B}$, and the resistive power loss $\mathbf{J} \cdot \mathbf{E}$) we see that they all represent bounded linear functionals which map

TABLE 1.2: Electromagnetic Quantities and Their Function Space

QUANTITY	FUNCTION SPACE
E, H	$\mathcal{H}(\text{Curl}; \Omega)$
D, B, J, M	$\mathcal{H}(\text{Div}; \Omega)$
ρ_e, ρ_m	$\mathcal{L}^2(\Omega)$
$\varepsilon_0 \bar{\bar{\varepsilon}}_r, \mu_0 \bar{\bar{\mu}}_r$	$\mathcal{H}(\text{Curl}; \Omega) \longrightarrow \mathcal{H}(\text{Div}; \Omega)$

either $\mathcal{H}(\text{Curl}; \Omega)$ or $\mathcal{H}(\text{Div}; \Omega)$ onto $\mathbb{C}$. Thus, we can say that $\mathcal{H}(\text{Curl}; \Omega)$ is dual to $\mathcal{H}(\text{Div}; \Omega)$ or that

$$\mathcal{H}(\text{Curl}; \Omega) = (\mathcal{H}(\text{Div}; \Omega))' \quad \text{and} \quad \mathcal{H}(\text{Div}; \Omega) = (\mathcal{H}(\text{Curl}; \Omega))'. \qquad (1.45)$$

1.2   PARAMETRIC GEOMETRY FUNDAMENTALS

In the next section, we will discuss classes of discrete or finite elements (used for the discretization of the field quantities) that conform to the vector spaces obeyed by the electromagnetic quantities. These finite elements consist of localized domains over which vector functions can be defined. That is, we consider the discretized domain $\Omega_h \subset \Omega$ and partitions of Ω_h denoted by the set of subdomains $\{\Omega_n\}$ such that $\Omega_h = \bigcup_{n=1}^{N} \Omega_n$, where $\Omega_n \cap \Omega_m = \{\emptyset\}$ for $n \neq m$. The discretized domain Ω_h is parameterized by the maximum diameter h of the subdomains Ω_n, and we shall assume that as $h \to 0$, $\Omega_h \to \Omega$.

The two most ubiquitous volume elements used to partition domain Ω_h in three-dimensional (3D) space are the tetrahedral and hexahedral finite elements. Each element type has its own advantages. Specifically, tetrahedrals form a simplex (i.e., a tetrahedral is the minimum volume containing $3 + 1$ nodes [30]) and can be used to partition practically any electromagnetic geometry using Delaunay triangulation. Also, in some cases, analytic expressions can be found for some integral expressions over such elements. On the other hand, hexahedral elements fill Ω_h more efficiently than tetrahedral elements since they lead to a reduction in the number of unknowns needed to resolve electromagnetic quantities. To argue that one type of element is more desirable than the other is a rather controversial topic. In this book, we will mostly concentrate on using hexahedrals for 3D discretization since the authors have focused on these elements over the past few years. When using hexahedral finite elements, the discretized boundary $\partial\Omega_h$ is characterized by quadrilaterals, as opposed to triangular elements when tetrahedral volume elements are considered. Thus, in two-dimensional (2D) formulations, we will consider quadrilateral elements for the partitioning of Ω_h.

Because we are often concerned with curved bodies in the analysis of electromagnetic structures, we will also consider here second-order parametric finite elements [31–36]. These curvilinear finite elements are characterized by 27-node hexahedral elements, 9-node quadrilateral elements, and 3-node linear elements. Below, we will discuss the parametric transformations used in the description of these elements and define the necessary calculus over these elements that enable us to use them in the discretization of Maxwell's equations. For a description of tetrahedral elements and their higher order representations, the reader is referred to [37–39]. We note that even if the reader is not interested in using curvilinear elements, the idea of considering each element as a parameterized version of some standard element is very typical of most finite element implementations of Maxwell's equations. Further, all lower order finite elements considered in the literature can be thought of as special cases of the elements considered here. That is, rectilinear elements are first-order cases of curvilinear hexahedral and quadrilateral elements. Also, curvilinear tetrahedral and triangular elements as well as their low-order counterparts are actually degenerate forms of the elements considered in this book.

1.2.1   Parametric Geometry in 1D

Let us consider the 1D curvilinear element depicted in Fig. 1.1. In Chapter 2 this element will be used to describe the boundary within the context of a 2D finite element–boundary integral (FE–BI) formulation. For second-order parametric elements, we only need three nodes to fully define this element. The position vector can then be found everywhere along the surface of the element through the use of Lagrange interpolatory polynomials. Because we choose the parametric domain to be $[-1, 1]$, these interpolatory polynomials become

$$L_i(x) = \begin{cases} \frac{1}{2}x(x-1) & \text{if} \quad i = 0 \\ 1 - x^2 & \text{if} \quad i = 1 \\ \frac{1}{2}x(x+1) & \text{if} \quad i = 2. \end{cases} \tag{1.46}$$

Thus, the parametric transformation defining the position vector tracing the element is

$$\mathbf{r}(u) = \sum_{i=0}^{2} L_i(u)\mathbf{r}_i, \tag{1.47}$$

where $\mathbf{r}_i$ are the three nodal points. As can be observed, $L_i(u)$ is unity at the ith node and vanishes at the other nodes. Thus, (1.46) represents the weighted sum of the position vectors at the three nodes $\mathbf{r}_i$.

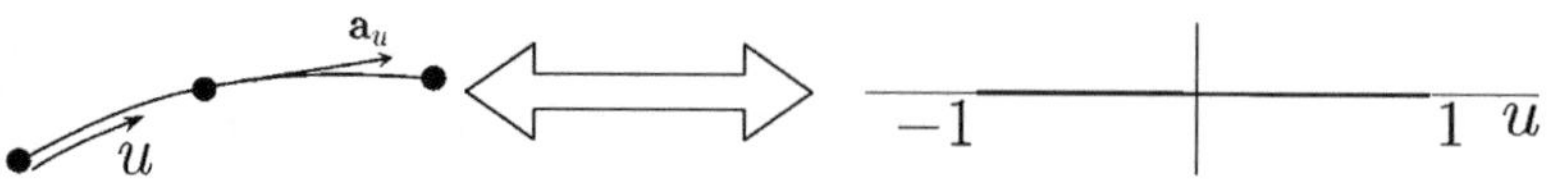

FIGURE 1.1: One-dimensional (1D) parametric transformation

We next proceed to introduce a set of vectors on the element. We call these *covariant* and *contravariant* unitary vectors. The covariant unitary vector is the most typical, given by

$$\mathbf{a}_u(u) = \lim_{\Delta u \to 0} \frac{\mathbf{r}(u + \Delta u) - \mathbf{r}(u)}{\Delta u} = \frac{\partial \mathbf{r}}{\partial u}. \qquad (1.48)$$

The inherent property of the covariant vector is that it is tangential to the element. However, because it is not a unit vector, unlike the $\hat{x}$ coordinate vector for Euclidean geometries, we must also define a contravariant unitary vector, denoted by $\mathbf{a}^u$, such that

$$\mathbf{a}^u \cdot \mathbf{a}_u = 1. \qquad (1.49)$$

It is then easily seen that the contravariant unitary vector must have the form

$$\mathbf{a}^u(u) = \frac{1}{|\mathbf{a}_u|^2} \mathbf{a}_u(u). \qquad (1.50)$$

It is of course necessary to also define differential operators on these elements, i.e., gradient and divergence operators (note that the curl operator is undefined in 1D). The gradient of a scalar function, $\phi(u)$, points in the direction of the most rapid change and has magnitude equal to its rate of change in that direction. Thus, it should seem plausible that the gradient of $\phi(u)$ be given by

$$\nabla \phi(u) = \frac{\partial \phi}{\partial u} \mathbf{a}^u, \qquad (1.51)$$

where the rate of change of $\nabla \phi(u)$ in the direction $\mathbf{a}_u$ is $\nabla \phi(u) \cdot \mathbf{a}_u = \frac{\partial \phi}{\partial u}$. Additionally, the divergence of a vector function, $\mathbf{f}(u)$, in 1D is given by

$$\nabla \cdot \mathbf{f}(u) = \frac{1}{|\mathbf{a}_u|^2} \left(\frac{\partial [(\mathbf{f} \cdot \mathbf{a}^u)|\mathbf{a}_u|^2]}{\partial u} \right). \qquad (1.52)$$

We defer the explanation of this derivation to the subsequent section (2D definition of the divergence operator).

In this work, one-dimensional analysis is needed to calculate boundary integrals or line integrals in two-dimensional space. Because these integrals are evaluated in parametric space, we need to perform a variable transformation in which the infinitesimal line integral quantity dl is given by

$$dl = |\mathbf{a}_u| du. \qquad (1.53)$$

Here term $|\mathbf{a}_u|$ is most often referred to as the *Jacobian* of the variable transformation.

1.2.2 Parametric Geometry in 2D

We now consider the 9-node quadrilateral finite element depicted in Fig. 1.2. The position vector everywhere on the surface of this element is parameterized by the second-order Lagrange interpolatory polynomials defined in (1.46) and is given by

$$\mathbf{r}(u, v) = \sum_{i=0}^{2} \sum_{j=0}^{2} L_i(u) L_j(v) \mathbf{r}_{ij}, \qquad (1.54)$$

where $\mathbf{r}_{ij}$ defines the nine nodes of the curvilinear quadrilateral element.

As was done for the 1D case, we proceed to define a set of covariant and contravariant coordinate vectors. The *covariant* coordinate vectors, denoted as $\mathbf{a}_u$ and $\mathbf{a}_v$, are defined, respectively, as

$$\mathbf{a}_u(u, v) = \lim_{\Delta u \to 0} \frac{\mathbf{r}(u + \Delta u, v) - \mathbf{r}(u, v)}{\Delta u} = \frac{\partial \mathbf{r}}{\partial u}, \qquad (1.55)$$

and

$$\mathbf{a}_v(u, v) = \lim_{\Delta v \to 0} \frac{\mathbf{r}(u, v + \Delta v) - \mathbf{r}(u, v)}{\Delta v} = \frac{\partial \mathbf{r}}{\partial v}. \qquad (1.56)$$

Note that $\mathbf{a}_u$ and $\mathbf{a}_v$ are both tangent to the element surface (see Fig. 1.2) and point in the direction of constant v and u, respectively. These coordinate vectors are not unit vectors, implying that their lengths will scale according to the physical size of the patch. Also, these coordinate vectors need not be perpendicular to each other (see Fig. 1.3), albeit independent. Therefore, to complete the definition of the 2D parametric space, an additional set of vectors

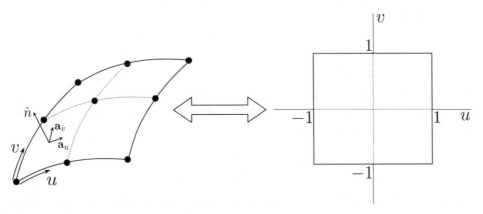

FIGURE 1.2: Two-dimensional parametric transformation

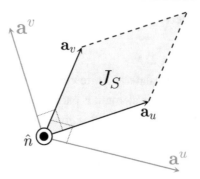

FIGURE 1.3: Two-dimensional covariant–contravariant vector relationship

are needed, namely the *contravariant* coordinate vectors $\mathbf{a}^u$ and $\mathbf{a}^v$ defined as

$$\mathbf{a}^u(u, v) = \frac{1}{J_S}\mathbf{a}_v \times \hat{n},\tag{1.57}$$

and

$$\mathbf{a}^v(u, v) = \frac{1}{J_S}\hat{n} \times \mathbf{a}_u.\tag{1.58}$$

Here the surface Jacobian is given by $J_S = |\mathbf{a}_u \times \mathbf{a}_v|$ with the unit normal to the surface given by $\hat{n} = (\mathbf{a}_u \times \mathbf{a}_v)/J_S$. As expected, the surface Jacobian is used to perform the variable transformation within surface integrals, e.g.,

$$dS = J_S \, dudv.\tag{1.59}$$

Similar to the covariant unitary vectors, the contravariant unitary vectors are tangent to the patch surface. However, they do not in general point in the direction of constant u or v. Further, the contravariant vectors are not necessarily unit vectors, having lengths inverse to those of the covariant vector, so that

$$\mathbf{a}_i \cdot \mathbf{a}^j = \delta_{ij} \qquad \text{for} \quad i, j = u, v,\tag{1.60}$$

where δ_{ij} is the Kronecker delta-function.

We can now proceed to explicitly define the differential operators (e.g., gradient, divergence, and curl). The gradient of a scalar function, $\phi(u, v)$, can be derived using similar arguments made for the 1D case, namely,

$$\nabla\phi(u, v) = \frac{\partial\phi}{\partial u}\mathbf{a}^u + \frac{\partial\phi}{\partial v}\mathbf{a}^v.\tag{1.61}$$

In general, the divergence of a vector in two dimensions (often called the surface divergence) is defined as the ratio of net flux leaving an infinitesimally small curve encircling the point of

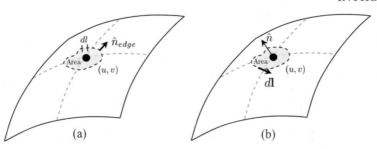

FIGURE 1.4: Schematic to define (a) divergence and (b) curl on a curvilinear surface

interest to the surface's area (see Fig. 1.4(a)). In mathematical terms, this definition is equivalent to the expression

$$\nabla \cdot \mathbf{f}(u, v) = \lim_{\text{Area} \to 0} \frac{\oint \hat{n}_{\text{edge}} \cdot \mathbf{f} dl}{\text{Area}}. \tag{1.62}$$

Note that $\hat{n}_{\text{edge}}$ is not the normal to the patch's surface but is normal to the contour bounding the surface patch. From this definition and the concepts presented above, the divergence can be represented as

$$\nabla \cdot \mathbf{f}(u, v) = \lim_{J_S \Delta u \Delta v \to 0} \frac{\mathbf{f}_u J_S|_{u+\Delta u, v} \Delta v + \mathbf{f}_v J_S|_{u, v+\Delta v} \Delta u - \mathbf{f}_u J_S|_{u, v} \Delta v - \mathbf{f}_v J_S|_{u, v} \Delta u}{J_S \Delta u \Delta v}, \tag{1.63}$$

where $\mathbf{f}_i = \mathbf{f} \cdot \mathbf{a}^i$. A more compact form of (1.63) can be written as

$$\nabla \cdot \mathbf{f}(u, v) = \frac{1}{J_S} \left(\frac{\partial [(\mathbf{f} \cdot \mathbf{a}^u) J_S]}{\partial u} + \frac{\partial [(\mathbf{f} \cdot \mathbf{a}^v) J_S]}{\partial v} \right). \tag{1.64}$$

An argument similar to the above holds for the derivation of the curl operator in 2D. In general, the curl is defined as the ratio of the circulation around an infinitesimally small curve encircling the point of interest to the surface's area (see Fig. 1.4(b)). This definition is equivalent to the expression

$$\nabla \times \mathbf{f}(u, v) = \lim_{\text{Area} \to 0} \frac{\oint \mathbf{f} \cdot d\mathbf{l}}{\text{Area}} \hat{n}. \tag{1.65}$$

Thus, the curl can be shown to have the form

$$\nabla \times \mathbf{f}(u, v) = \lim_{J_S \Delta u \Delta v \to 0} \frac{\mathbf{f}^v|_{u+\Delta u, v} \Delta v + \mathbf{f}^u|_{u, v+\Delta v} \Delta u - \mathbf{f}^v|_{u, v} \Delta v - \mathbf{f}^u|_{u, v} \Delta u}{J_S \Delta u \Delta v} \hat{n}, \tag{1.66}$$

where $\mathbf{f}^i = \mathbf{f} \cdot \mathbf{a}_i$. Casting this into a more compact form leads to

$$\nabla \times \mathbf{f}(u, v) = \frac{1}{J_S} \left[\frac{\partial (\mathbf{f} \cdot \mathbf{a}_v)}{\partial u} - \frac{\partial (\mathbf{f} \cdot \mathbf{a}_u)}{\partial v} \right] \hat{n}. \tag{1.67}$$

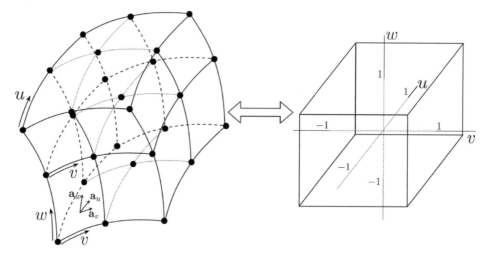

FIGURE 1.5: Parametric transformation for hexahedrals

1.2.3   Parametric Geometry in 3D

The extension of the prescribed definitions and concepts to three dimensions is rather straight-forward. Therefore, a complete derivation will be omitted here and the expressions required for curvilinear hexahedral finite elements will simply be stated.

The position vector anywhere inside the hexahedral element (see Fig. 1.5) takes the form

$$\mathbf{r}(u, v, w) = \sum_{i=0}^{2}\sum_{j=0}^{2}\sum_{k=0}^{2} L_i(u)L_j(v)L_k(w)\mathbf{r}_{ijk}, \qquad (1.68)$$

where (u, v, w) are associated with the coordinates shown in Fig. 1.5. Also, $\mathbf{r}_{ijk}$ denotes the 27 nodal points defining the hexahedra. A set of covariant coordinate vectors are defined as

$$\mathbf{a}_u = \frac{\partial \mathbf{r}}{\partial u}, \qquad \mathbf{a}_v = \frac{\partial \mathbf{r}}{\partial v}, \qquad \text{and} \qquad \mathbf{a}_w = \frac{\partial \mathbf{r}}{\partial w}, \qquad (1.69)$$

with the corresponding set of contravariant coordinate vectors given by

$$\mathbf{a}^u = \frac{1}{J_V}(\mathbf{a}_v \times \mathbf{a}_w), \qquad \mathbf{a}^v = \frac{1}{J_V}(\mathbf{a}_w \times \mathbf{a}_u), \qquad \text{and} \qquad \mathbf{a}^w = \frac{1}{J_V}(\mathbf{a}_u \times \mathbf{a}_v). \qquad (1.70)$$

Here the Jacobian of the volumetric transformation is given by $J_V = \mathbf{a}_u \cdot (\mathbf{a}_v \times \mathbf{a}_w)$ (see Fig. 1.6) and $dV = J_V\,dudvdw$. The gradient, divergence, and curl operators can all be derived following the same procedures used for their 2D counterparts and are simply stated here,

$$\nabla\phi = \frac{\partial\phi}{\partial u}\mathbf{a}^u + \frac{\partial\phi}{\partial v}\mathbf{a}^v + \frac{\partial\phi}{\partial w}\mathbf{a}^w, \qquad (1.71)$$

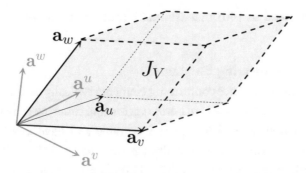

FIGURE 1.6: Three-dimensional graphical display of the covariant and contravariant vectors

$$\nabla \cdot \mathbf{f} = \frac{1}{J_V}[\frac{\partial(\mathbf{f}_u J_V)}{\partial u} + \frac{\partial(\mathbf{f}_v J_V)}{\partial v} + \frac{\partial(\mathbf{f}_w J_V)}{\partial w}], \qquad (1.72)$$

$$\nabla \times \mathbf{f} = \frac{1}{J_V}[(\frac{\partial \mathbf{f}^w}{\partial v} - \frac{\partial \mathbf{f}^v}{\partial w})\mathbf{a}_u + (\frac{\partial \mathbf{f}^u}{\partial w} - \frac{\partial \mathbf{f}^w}{\partial u})\mathbf{a}_v + (\frac{\partial \mathbf{f}^v}{\partial u} - \frac{\partial \mathbf{f}^u}{\partial v})\mathbf{a}_w], \qquad (1.73)$$

where again $\mathbf{f}^i = \mathbf{f} \cdot \mathbf{a}^i$ and $\mathbf{f}_i = \mathbf{f} \cdot \mathbf{a}_i$ for $i \in \{u, v, w\}$.

1.3 CURVILINEAR FINITE ELEMENTS

For the hybrid 2D and 3D computational electromagnetic formulations to be described in this text, we are required to develop finite elements in $\mathcal{H}(\text{Curl}; \Omega)$ to represent field intensities. Concurrently, we must employ boundary elements in $\mathcal{H}^{-1/2}(\text{Div}; \partial\Omega)$ to represent equivalent surface currents. With this in mind, in this section, we make use of the parametric geometry framework defined above to derive finite elements that either conform to the space $\mathcal{H}(\text{Curl}; \Omega)$ or $\mathcal{H}^{-1/2}(\text{Div}; \partial\Omega)$. However, we mention here that once finite elements are constructed for $\mathcal{H}(\text{Curl}; \Omega)$, the associated boundary elements are valid in $\mathcal{H}^{-1/2}(\text{Div}; \partial\Omega)$ and $\mathcal{H}^{-1/2}(\text{Curl}; \partial\Omega)$ and this follows from their definitions (1.30) and (1.33) where the trace operators $\gamma_\times$ and γ_t are applied to vector functions in $\mathcal{H}(\text{Curl}; \Omega)$, respectively.

We say that a set of vector functions are *curl conforming* if they span $\mathcal{H}(\text{Curl}; \Omega_h)$ and are a subset of $\mathcal{H}(\text{Curl}; \Omega)$ for any value of h (i.e., any mesh density). Similarly, we say that a set of vector functions are *divergence conforming* if they span $\mathcal{H}^{-1/2}(\text{Div}; \partial\Omega_h)$ and are a subset of $\mathcal{H}^{-1/2}(\text{Div}; \partial\Omega)$ for any value of h. As a side note, the well-known RWG (Rao–Wilton–Glisson) boundary elements defined over triangular meshes [7] are contained in the space $\mathcal{H}^{-1/2}(\text{Div}; \partial\Omega_h)$. In this section, we consider a class of divergence conforming boundary elements used in 2D and 3D formulations, first the rooftop counterparts to the RWG boundary elements. We then describe curl conforming finite elements in 2D and 3D which are parametric versions of those proposed in [40] that eliminate spurious solutions of Maxwell's equations [41, 42].

1.3.1 Divergence Conforming Elements

We begin by considering divergence conforming elements typically used to discretize equivalent surface currents. Essentially, the task is to define vector functions over the finite element in Fig. 1.1. These vector functions should be square integrable and have a well-defined divergence over the entire boundary element. Additionally, the divergence itself must be square integrable over the whole boundary element domain $\partial\Omega_b$. Though not yet clear, the divergence theorem requires that the component normal to the element edges be continuous (for 2D boundary elements, these edges correspond to the two end nodes in Fig. 1.1 and the vector $\mathbf{a}_u$ is always normal to these edges/nodes). Thus, let us consider the following two vector functions over the finite element (line segment) in Fig. 1.1:

$$\mathbf{v}_1^{2D} = (1 + u)\frac{1}{|\mathbf{a}_u|^2}\mathbf{a}_u \quad \text{and} \quad \mathbf{v}_2^{2D} = (1 - u)\frac{1}{|\mathbf{a}_u|^2}\mathbf{a}_u. \tag{1.74}$$

The divergence of these two vector functions using (1.52) is $\pm\frac{1}{|\mathbf{a}_u|^2}$. In addition, we see that when $u = -1$ then $\mathbf{v}_1^{2D} = 0$ and when $u = 1$ then $\mathbf{v}_2^{2D} = 0$. This property allows both $\mathbf{v}_1$ and $\mathbf{v}_2$ to be paired with vector functions defined on adjacent elements to form triangular-shaped basis functions as seen in Fig. 1.7. The necessity of this pairing is stressed in Fig. 1.8 where we consider the derivative of a function with and without continuity at element–element junctions. Also note that the capability of pairing these vector functions with those from adjacent elements ensures the square integrable divergence property of $\mathcal{H}^{-1/2}(\text{Div}; \partial\Omega_b)$ throughout the support $\partial\Omega_b$ of the boundary element.

We next consider divergence conforming vector functions defined over a curvilinear quadrilateral finite element depicted in Fig. 1.2. These elements will be used as (surface)

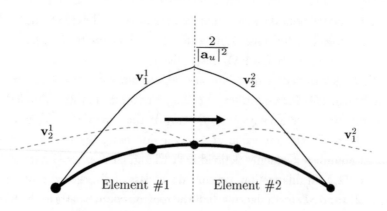

FIGURE 1.7: 2D divergence conforming boundary elements

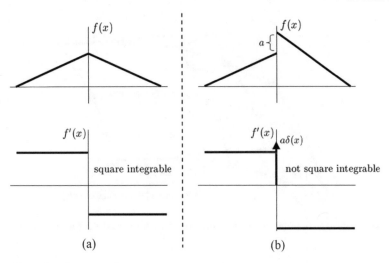

FIGURE 1.8: Necessity for normal component continuity

boundary elements in 3D hybrid formulations. The properties of the vector functions defined over these elements are similar to those described above for the 2D case and are given by

$$
\begin{aligned}
\mathbf{v}_1^{3D} &= (1+u)\frac{1}{J_S}\mathbf{a}_u \quad \text{and} \quad \mathbf{v}_2^{3D} = (1-u)\frac{1}{J_S}\mathbf{a}_u, \\
\mathbf{v}_3^{3D} &= (1+v)\frac{1}{J_S}\mathbf{a}_v \quad \text{and} \quad \mathbf{v}_4^{3D} = (1-v)\frac{1}{J_S}\mathbf{a}_v.
\end{aligned}
\tag{1.75}
$$

It is a rather simple task to show that the divergence (1.64) of each vector function is well defined. In addition, each one of these vector functions is typically associated with an edge of the quadrilateral element as shown in Fig. 1.2. Thus, as displayed in Fig. 1.9 each vector function can be paired with a corresponding one from an adjacent element to ensure normal component continuity. Because of the obvious rooftop resemblance of these shape functions, they are often referred to as *rooftop* divergence conforming vector functions.

1.3.2 Curl Conforming Elements

The most important finite elements in computational electromagnetics are the curl conforming elements used to discretize electric and magnetic field intensities for solutions of Maxwell's equations. In a similar way to the boundary elements described above, the vector functions defined over these elements must be square integrable but require that their curl also be square integrable throughout Ω_h. Due to Stoke's theorem, it is possible to show that this is guaranteed as long as the components of the vector functions tangential to the elements' edges are continuous. In this manner, we eliminate spurious solutions arising from the improper modeling of the null

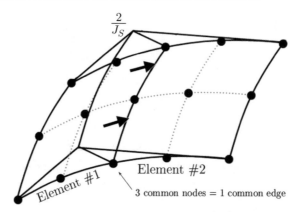

FIGURE 1.9: 3D rooftop divergence conforming (ensuring normal vector field continuity) boundary element

space in PDE-based solutions of Maxwell's vector wave equation [31, 40, 43]. Next, we first propose curl conforming elements valid for 2D formulations. In this context, we consider the four vector functions defined over the finite element in Fig. 1.2, namely,

$$
\begin{aligned}
\mathbf{w}_1^{2D} &= (1 + v)\mathbf{a}^u \quad &&\text{and} \quad &&\mathbf{w}_2^{2D} = (1 - v)\mathbf{a}^u, \\
\mathbf{w}_3^{2D} &= (1 + u)\mathbf{a}^v \quad &&\text{and} \quad &&\mathbf{w}_4^{2D} = (1 - u)\mathbf{a}^v.
\end{aligned}
\tag{1.76}
$$

The curl of these four vector functions using (1.67) gives the result $\pm\frac{1}{J_s}\hat{n}$. In addition, each vector function is associated with an edge of the finite element depicted in Fig. 1.2, and can therefore be paired with a vector function from an adjacent element (see Fig. 1.10). Interestingly, the ability to enforce tangential component continuity across element–element interfaces allows for natural satisfaction of the boundary conditions associated with electric and magnetic fields.

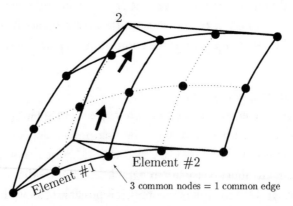

FIGURE 1.10: 2D curl conforming (ensuring tangential vector field continuity) finite element

We now proceed to consider curl conforming vector functions defined over the curvilinear hexahedral finite element depicted in Fig. 1.5. These elements will be used in 3D formulations. The properties of the vector functions defined over these elements are a generalization to those described for the 2D case and are given by

$$
\begin{aligned}
\mathbf{w}_1^{3D} &= (1+v)(1+w)\mathbf{a}^u & \mathbf{w}_5^{3D} &= (1+u)(1+w)\mathbf{a}^v & \mathbf{w}_9^{3D} &= (1+u)(1+v)\mathbf{a}^w, \\
\mathbf{w}_2^{3D} &= (1+v)(1-w)\mathbf{a}^u & \mathbf{w}_6^{3D} &= (1+u)(1-w)\mathbf{a}^v & \mathbf{w}_{10}^{3D} &= (1+u)(1-v)\mathbf{a}^w, \\
\mathbf{w}_3^{3D} &= (1-v)(1+w)\mathbf{a}^u & \mathbf{w}_7^{3D} &= (1-u)(1+w)\mathbf{a}^v & \mathbf{w}_{11}^{3D} &= (1-u)(1+v)\mathbf{a}^w, \\
\mathbf{w}_4^{3D} &= (1-v)(1-w)\mathbf{a}^u & \mathbf{w}_8^{3D} &= (1-u)(1-w)\mathbf{a}^v & \mathbf{w}_{12}^{3D} &= (1-u)(1-v)\mathbf{a}^w.
\end{aligned}
$$

$$(1.77)$$

It is a rather simple task to show that the curl (1.73) of each vector function is well defined. In addition, each one of these vector functions are associated with one of the 12 edges of the hexahedral element in Fig. 1.5. Similar to the 2D case, we see that each vector function (see Fig. 1.11) can be paired with a corresponding vector function from adjacent elements to ensure the necessary tangential component continuity across element interfaces.

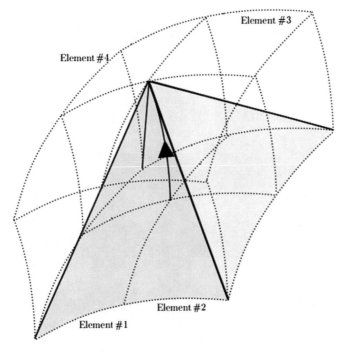

FIGURE 1.11: 3D curl conforming finite element

1.4 OVERVIEW

We presented, in this chapter, a brief theoretical foundation of the numerical solution of partial differential equations (PDEs) and integral equations (IEs). As a result, a representation of Sobolev spaces was presented and the related implications for a unique solution of the ensuing PDEs and IEs were discussed. Among the implications were critical restrictions related to basis functions used for field expansion as well as the testing functions used in the context of Galerkin and Petrov–Galerkin formulations, which lead to discrete systems. We proceeded then to introduce appropriate basis/expansion functions for 2D and 3D applications. These were based on quadrilateral and hexahedral element discretization, and the reader is referred to other references and texts for tetrahedrals and triangular discretizations [32, 37–40].

CHAPTER 2

Two-Dimensional Hybrid FE–BI

The finite element method (FEM) has been demonstrated as one of the most efficient and versatile computational tools [37], especially when considering closed-domain structures such as waveguides [44]. However, when radiation and/or scattering applications are considered, where the domain of interest is infinite (i.e., an open-domain problem), we must (for the sake of finite computational resources) implement a domain truncation scheme. One such scheme involves the application of absorbing boundary conditions, or perfectly matched layers (PMLs), around the object of interest [37]. Due to the approximate nature of absorbing boundary conditions, it is often more desirable to truncate the FEM domain using a boundary integral, a technique first introduced to the electromagnetics community by Silvester in [45] and by McDonald in [46]. We call this technique the hybrid finite element–boundary integral (FE–BI) technique. However, because of the required computational resources the FE–BI was not applied and developed for three-dimensional (3D) applications untill the late 1980s and early 1990s. The introduction of edge elements [40] coupled with fast iterative solvers such as the CGFFT [47, 48] provided the basis for the success of FE–BI for solving 3D EM scattering and radiation problems. Nearly 10 years after the first full 3D implementation, the FE–BI methods are now among the mainstream methods for EM analysis. The FE–BI technique exploits the surface equivalence principle [28] to decouple the interior and boundary fields of the FEM domain. In other words, the fields are expressed everywhere on the surface of the FEM domain using a boundary integral characterized by equivalent sources representing both the tangential electric and magnetic field components. On the other hand, the fields inside the FEM domain are formulated using a discretization of the PDE equation. Coupling of the interior and exterior fields is enforced by applying the appropriate continuity across the boundaries. As a result of this hybrid approach, we will see in this chapter the efficiency of the FE–BI to simulate the scattering from very complex penetrable structures consisting of, but not limited to, inhomogeneous anisotropic materials.

Because all aspects of a 3D FE–BI can be discussed in the context of a two-dimensional (2D) FE–BI, we consider (in this chapter) the 2D FE–BI formulation for scattering by penetrable cylinders (see Fig. 2.1). Previous works on 2D FE–BI can be found in [49–60]. Though this list is surely not meant to be comprehensive, it should give the reader a good overview of the previous work. As noted, most of the early work on the 2D FE–BI was done in the late 1980s to early 1990s. However, in light of the recent work [22] (where a symmetric FE–BI formulation for 3D structures was implemented based upon the works in [61, 62]), we will update previous 2D FE–BI formulations. We will do this by adhering to mathematical principles developed in Chapter 1. The resulting formulation does contain internal resonance problems [63–71], though the resulting effects do not appear to be as significant as in the previous formulations [22], especially when diagonal scaling is performed [72].

We begin below by developing 2D wave equations for two special cases of transverse electric (TE) and transverse magnetic (TM) polarizations. We then complete the BVP statement by discussing all possible boundary conditions, both natural and approximate, satisfied for each polarization. The variational formulation of the BVP is then constructed and the incorporation of boundary integrals is accomplished so that a symmetric system matrix is generated. Discretization schemes are further presented along with results demonstrating the capability of the proposed method. Overall, it is the goal of this chapter to aid the reader in the field of computational electromagnetics by presenting a general, up-to-date FE–BI formulation.

2.1   THE BOUNDARY VALUE PROBLEM

We develop here the 2D FE–BI formulation to model the scattering from a cylinder of arbitrary shape and composition as shown in Fig. 2.1. In general, the cylinder can be composed of anisotropic material regions consisting of perfect electric conducting (PEC), perfect magnetic conducting (PMC), impedance, and/or material sheet surfaces. Although no material inhomogeneity is assumed in the longitudinal $\hat{z}$-direction of the cylinder, we do allow for the existence

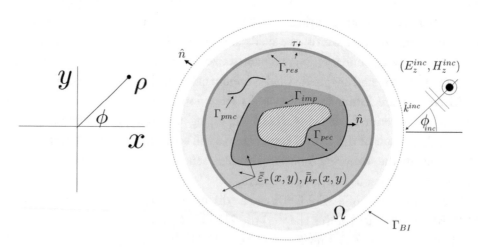

FIGURE 2.1: Geometry of a 2D cylinder of arbitrary shape and composition

of inhomogeneities in the transverse plane (xy-plane). In addition, a plane wave directed within the transverse plane is assumed to be incident on the structure. This enables us to develop the FE–BI formulation for the two cases where either the electric field $\mathbf{E}$ is completely transverse to $\hat{z}$ (i.e., $\hat{z} \cdot \mathbf{E} = 0$) or the magnetic field $\mathbf{H}$ is completely transverse to $\hat{z}$. These two types of polarizations are most commonly referred to as the TE and TM polarizations (or E_z and H_z incidences), respectively. We note that for the TE polarization the magnetic field is completely polarized in the $\hat{z}$-direction (i.e., $\hat{z} \times \mathbf{H} = 0$), and similarly $\hat{z} \times \mathbf{E} = 0$ for the TM polarization.

2.1.1   TE Polarization

Assuming a source-free material region characterized by $\bar{\bar{\varepsilon}}_r$ and $\bar{\bar{\mu}}_r$, we begin the FE–BI formulation by considering the following time-harmonic vector wave equation derived from (1.39) and (1.40),

$$\nabla \times \bar{\bar{\mu}}_r^{-1} \cdot \nabla \times \mathbf{E} - k_0^2 \bar{\bar{\varepsilon}}_r \cdot \mathbf{E} = 0, \tag{2.1}$$

where k_0 is the free-space wavenumber with the relative anisotropic permittivity and permeability dyads given by

$$\bar{\bar{\varepsilon}}_r(x, y) = \begin{bmatrix} \varepsilon_{xx} & \varepsilon_{xy} & 0 \\ \varepsilon_{yx} & \varepsilon_{yy} & 0 \\ 0 & 0 & \varepsilon_z \end{bmatrix} \quad \text{and} \quad \bar{\bar{\mu}}_r(x, y) = \begin{bmatrix} \mu_{xx} & \mu_{xy} & 0 \\ \mu_{yx} & \mu_{yy} & 0 \\ 0 & 0 & \mu_z \end{bmatrix}. \tag{2.2}$$

In general, it is possible to express the electric field as a combination of its transverse and longitudinal components,

$$\begin{aligned} \mathbf{E} &= \hat{z} \times (\mathbf{E} \times \hat{z}) + E_z \hat{z} \\ &= \mathbf{E}_t + E_z \hat{z}. \end{aligned} \tag{2.3}$$

Similarly, it is helpful to express the ∇ operator as

$$\begin{aligned} \nabla &= \hat{x}\frac{\partial}{\partial x} + \hat{y}\frac{\partial}{\partial y} + \hat{z}\frac{\partial}{\partial z} \\ &= \nabla_t + \hat{z}\frac{\partial}{\partial z}. \end{aligned} \tag{2.4}$$

However, for TE polarization (i.e., $E_z = 0$), $\mathbf{E} = \mathbf{E}_t$ and thus (2.1) becomes

$$\nabla_t \times \frac{1}{\mu_z} \nabla_t \times \mathbf{E}_t - k_0^2 \bar{\bar{\varepsilon}}_r \cdot \mathbf{E}_t = 0. \tag{2.5}$$

In arriving at wave equation (2.5) we have exploited the fact that the electric field possesses no z-dependence, a property resulting from our choice of excitation.

2.1.2   TM Polarization

Similarly, the derivation of the 2D PDE for TM polarized waves begins with the time-harmonic vector wave equation (for magnetic fields)

$$\nabla \times \bar{\bar{\varepsilon}}_r^{-1} \cdot \nabla \times \tilde{\mathbf{H}} - k_0^2 \bar{\bar{\mu}}_r \cdot \tilde{\mathbf{H}}_t = 0. \tag{2.6}$$

Here we have conveniently chosen $\tilde{\mathbf{H}}$ to be the augmented magnetic field intensity given by $\tilde{\mathbf{H}} = Z_0 \mathbf{H}$, where $Z_0 = \sqrt{\mu_0/\varepsilon_0}$ is the free-space impedance. Often, in hybrid formulations, we make use of this augmented field because it allows for all electromagnetic quantities to have the same units (volts in this case). The corresponding numerical solutions may also be better conditioned with this choice of unknowns.

As expected, for TM polarization $\tilde{\mathbf{H}} = \hat{z} \times (\tilde{\mathbf{H}} \times \hat{z}) = \tilde{\mathbf{H}}_t$. Thus, the wave equation in (2.6) can be succinctly written as

$$\nabla_t \times \frac{1}{\varepsilon_z} \nabla_t \times \tilde{\mathbf{H}}_t - k_0^2 \bar{\bar{\mu}}_r \cdot \tilde{\mathbf{H}}_t = 0, \tag{2.7}$$

where again we have exploited the fact that the magnetic field contains no z-dependence. Of course, (2.7) is the dual of (2.5) and this allows us to construct future expressions within the FE–BI formulation for TM polarization from the TE expressions.

2.1.3   Boundary Conditions

The PDEs in (2.5) and (2.7) have an infinite number of solutions. To arrive at a unique solution, we must explicitly describe the behavior of the fields at the boundaries of the domain. Together with the PDE in either (2.5) or (2.7), a complete boundary value problem (BVP) statement can then be formed. These so-called boundary conditions can be classified into two types, namely, the natural boundary conditions (NBCs) and the approximate boundary conditions (ABCs). Typically, the NBCs consist of the continuity conditions at material interfaces and conditions on PEC and PMC boundaries. The ABCs considered here are the impedance surface and resistive/conductive sheet boundary conditions. The reader is referred to [73] for an exhaustive discussion on ABCs in electromagnetics. In addition, because we are considering scattering in an open domain, the radiation boundary condition must always be enforced for all applications.

Radiation Boundary Condition

When the outer boundary domain recedes to infinity, as is the case in scattering and radiation problems, solutions to (2.5) and (2.7) must satisfy the Sommerfeld radiation condition

$$\lim_{\rho \to \infty} \sqrt{\rho} \left[\nabla_t \times \begin{pmatrix} \mathbf{E}_t \\ \tilde{\mathbf{H}}_t \end{pmatrix} + jk_0 \hat{\rho} \times \begin{pmatrix} \mathbf{E}_t \\ \tilde{\mathbf{H}}_t \end{pmatrix} \right] = 0, \tag{2.8}$$

to describe the field behavior at infinity in 2D for both TE and TM solutions with $\rho = \sqrt{x^2 + y^2}$. For plane wave incidence, (2.8) only applies to the scattered fields. This simply states that at infinity the electric and magnetic fields are outgoing and of the form $e^{-jk_0\rho}/\sqrt{\rho}$. We note that this pseudoboundary condition can be used as the lowest order absorbing boundary condition [73] when a truncated computational domain is applied to open domain problems. Because of the limited accuracy of absorbing boundary conditions, we present here a hybrid technique resulting from the truncation of the computational domain Ω by the boundary Γ_{BI} (see Fig. 2.1). In the process of doing so, for TE polarization, we introduce the augmented $(\tilde{\mathbf{J}}_s = Z_0 \mathbf{J}_s)$ electric surface current $\tilde{\mathbf{J}}_s = \hat{n} \times \tilde{\mathbf{H}}$ to represent the tangential magnetic field on Γ_{BI}. We may also state that

$$\hat{n} \times \frac{1}{\mu_z}(\nabla_t \times \mathbf{E}_t) = -jk_0\tilde{\mathbf{J}}_s \qquad (2.9)$$

on this boundary. Likewise, for TM polarization, we introduce the magnetic surface current $\mathbf{M}_s = \mathbf{E} \times \hat{n}$ to represent the tangential electric field on Γ_{BI}, implying

$$\hat{n} \times \frac{1}{\varepsilon_z}(\nabla_t \times \tilde{\mathbf{H}}_t) = -jk_0\mathbf{M}_s \qquad (2.10)$$

on Γ_{BI}.

Natural Boundary Conditions
At any boundary interface (see Fig. 2.2(a)) separating region I from region II, we enforce the tangential field continuity (or jump) relations

$$\begin{aligned}
\hat{n} \times (\mathbf{E}^{\mathrm{I}} - \mathbf{E}^{\mathrm{II}}) &= -\mathbf{M}_s \\
\hat{n} \times (\tilde{\mathbf{H}}^{\mathrm{I}} - \tilde{\mathbf{H}}^{\mathrm{II}}) &= \tilde{\mathbf{J}}_s \\
\hat{n} \cdot (\mathbf{D}^{\mathrm{I}} - \mathbf{D}^{\mathrm{II}}) &= \rho_{es} \\
\hat{n} \cdot (\mathbf{B}^{\mathrm{I}} - \mathbf{B}^{\mathrm{II}}) &= \rho_{ms},
\end{aligned} \qquad (2.11)$$

which can be derived from the integral form of Maxwell's equations [28]. Here $\mathbf{M}_s$ and $\tilde{\mathbf{J}}_s$ are the surface magnetic and electric currents at the boundary interface, and ρ_{es} and ρ_{ms} are the surface electric and magnetic charges, respectively.

For TE and TM polarizations, the material boundary conditions, describing the properties of the electric and magnetic fields at the boundary of a material discontinuity, can be stated after eliminating the surface currents and charges in (2.11) as

$$\begin{aligned}
\hat{n} \times \mathbf{E}_t^{\mathrm{I}} &= \hat{n} \times \mathbf{E}_t^{\mathrm{II}}, \\
\hat{n} \times \tilde{\mathbf{H}}_t^{\mathrm{I}} &= \hat{n} \times \tilde{\mathbf{H}}_t^{\mathrm{II}}.
\end{aligned} \qquad (2.12)$$

On surfaces of electrically conducting materials Γ_{pec}, the PEC boundary condition is characterized by the elimination of the tangential electric fields, namely,

$$\gamma_t \mathbf{E}_t|_{\Gamma_{\text{pec}}} = 0. \tag{2.13}$$

Similarly, on surfaces of magnetically conducting materials Γ_{pmc}, the PMC boundary condition is characterized by the vanishing tangential magnetic fields, namely,

$$\gamma_t \tilde{\mathbf{H}}_t|_{\Gamma_{\text{pmc}}} = 0. \tag{2.14}$$

Impedance Boundary Condition

The impedance boundary condition (IBC) is often used to model a thin coating as a simple single surface or simply to model an imperfect conductor. The IBC relates the electric and magnetic field components tangential to the impedance surface Γ_{imp} and is given here in its most general form as (see Fig. 2.2)

$$\gamma_t \mathbf{E} = \bar{\bar{\eta}} \cdot (\hat{n} \times \tilde{\mathbf{H}}). \tag{2.15}$$

For the cylinder in Fig. 2.1, the normalized surface impedance $\bar{\bar{\eta}}$ takes the form

$$\bar{\bar{\eta}} = \eta_{tt}\hat{t}\hat{t} + \eta_{zz}\hat{z}\hat{z}, \tag{2.16}$$

where $\hat{t}$ denotes the unit vector in the transverse plane, tangential to Γ_{imp}. For the case when the impedance boundary condition describes an imperfect conductor (see Fig. 2.2(b)) with lossy relative permittivity ε_r' and permeability μ_r', the normalized impedance is $\eta = \eta_{tt} = \eta_{zz} = \sqrt{\mu_r'/\varepsilon_r'}$ [73]. This boundary condition, though approximate, is rather accurate as long as the radius of curvature r_i of the surface is such that $|\text{Im}(\sqrt{\mu_r'\varepsilon_r'})|k_0 r_i \gg 1$ [37]. That is, we require the conducting region to be sufficiently smooth and lossy so that the penetrating fields do not reemerge at some other point. The impedance boundary condition is extensively used to model coated conductors (see Fig. 2.2(c)) having variable coating thickness τ. In this case the impedance can be computed using transmission line arguments. Specifically, it can be shown that the relative impedance on this boundary surface can be approximated as $\eta = j\sqrt{\mu_r'/\varepsilon_r'} \tan(k_0\sqrt{\mu_r'\varepsilon_r'}\tau)$ [73].

For TE polarization it is possible to rewrite (2.15) in terms of tangential electric field by substituting $\nabla_t \times \mathbf{E}_t = -jk_0\mu_z(\tilde{H}_z\hat{z})$ into (2.15) to get

$$\hat{n} \times \frac{1}{\mu_z}(\nabla_t \times \mathbf{E}_t) = -jk_0\frac{1}{\eta_{tt}}\gamma_t \mathbf{E}_t. \tag{2.17}$$

Likewise, for TM polarizations we can construct the dual of (2.17) to express (2.15) in terms of only the tangential magnetic field as

$$\hat{n} \times \frac{1}{\varepsilon_z}(\nabla_t \times \tilde{\mathbf{H}}_t) = -jk_0\eta_{zz}\gamma_t \tilde{\mathbf{H}}_t. \tag{2.18}$$

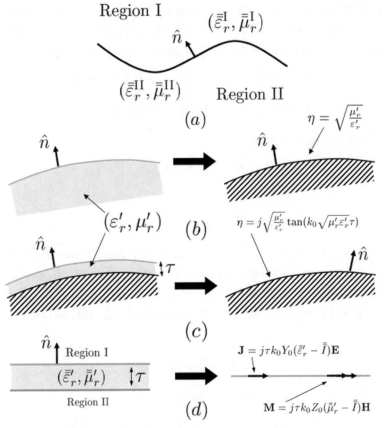

FIGURE 2.2: Boundary interfaces: (a) material discontinuity, (b) imperfect lossy conductor, (c) coated conductor, (d) composite sheet

Sheet Boundary Condition

The sheet boundary conditions are used to model thin material layers (see Fig. 2.2(d)). They are derived by considering the polarization currents induced within a thin material sheet given by $\mathbf{J} = jk_0 Y_0(\bar{\bar{\varepsilon}}_r' - \bar{\bar{I}}) \cdot \mathbf{E}_t$ and $\mathbf{M} = jk_0 Z_0(\bar{\bar{\mu}}_r' - \bar{\bar{I}}) \cdot \mathbf{H}_t$ (note that $Y_0 = 1/Z_0$). For our purposes, we note that both $\bar{\bar{\varepsilon}}_r'$ and $\bar{\bar{\mu}}_r'$ have the similar form as in (2.16). However, because the sheet of thickness τ is assumed to be sufficiently thin, $\mathbf{J}_s = \tau \gamma_t \mathbf{J}$ and $\mathbf{M}_s = \tau \gamma_t \mathbf{M}$. Then from (2.11) we have

$$\gamma_t(\mathbf{E}^{\mathrm{I}} + \mathbf{E}^{\mathrm{II}}) = 2\bar{\bar{R}}_e \cdot \hat{n} \times (\tilde{\mathbf{H}}^{\mathrm{I}} - \tilde{\mathbf{H}}^{\mathrm{II}})$$
$$\gamma_t(\tilde{\mathbf{H}}^{\mathrm{I}} + \tilde{\mathbf{H}}^{\mathrm{II}}) = -2\bar{\bar{R}}_m \cdot \hat{n} \times (\mathbf{E}^{\mathrm{I}} - \mathbf{E}^{\mathrm{II}}),$$

(2.19)

where the relative electric resistivity $\bar{\bar{R}}_e$ is given by $\bar{\bar{R}}_e = \frac{1}{jk_0\tau}(\bar{\bar{\varepsilon}}_r' - \bar{\bar{I}})^{-1}$ and the relative magnetic conductivity $\bar{\bar{R}}_m$ is given by $\bar{\bar{R}}_m = \frac{1}{jk_0\tau}(\bar{\bar{\mu}}_r' - \bar{\bar{I}})^{-1}$. As usual $(\mathbf{E}^{\mathrm{I}}, \tilde{\mathbf{H}}^{\mathrm{I}})$ and $(\mathbf{E}^{\mathrm{II}}, \tilde{\mathbf{H}}^{\mathrm{II}})$ refer to the fields above and below the sheet surface, respectively. Note that if $\bar{\bar{\varepsilon}}'_r = \bar{\bar{I}}$, then $\bar{\bar{R}}_e$ is undefined and $\hat{n} \times \tilde{\mathbf{H}}^{\mathrm{I}} = \hat{n} \times \tilde{\mathbf{H}}^{\mathrm{II}}$. Likewise, if $\bar{\bar{\mu}}'_r = \bar{\bar{I}}$, then $\bar{\bar{R}}_m$ is undefined and $\hat{n} \times \mathbf{E}^{\mathrm{I}} = \hat{n} \times \mathbf{E}^{\mathrm{II}}$.

For TE polarization, it is possible [74] to rewrite (2.19) using the matrix form

$$
\begin{bmatrix} \gamma_\times \mathbf{E}_t^{\mathrm{II}} \\ \gamma_t \tilde{\mathbf{H}}_t^{\mathrm{II}} \end{bmatrix} = \frac{1}{\Delta^{\mathrm{TE}}} \begin{bmatrix} \Lambda_{11}^{\mathrm{TE}} & \Lambda_{12}^{\mathrm{TE}} \\ \Lambda_{21}^{\mathrm{TE}} & \Lambda_{22}^{\mathrm{TE}} \end{bmatrix} \begin{bmatrix} \gamma_\times \mathbf{E}_t^{\mathrm{I}} \\ \gamma_t \tilde{\mathbf{H}}_t^{\mathrm{I}} \end{bmatrix},
\tag{2.20}
$$

where $\Delta^{\mathrm{TE}} = [4(\bar{\bar{R}}_e)_{tt}(\bar{\bar{R}}_m)_{zz} - 1]$ and

$$
\begin{aligned}
\Lambda_{11}^{\mathrm{TE}} &= \Lambda_{22}^{\mathrm{TE}} = [4(\bar{\bar{R}}_e)_{tt}(\bar{\bar{R}}_m)_{zz} + 1], \\
\Lambda_{12}^{\mathrm{TE}} &= 4(\bar{\bar{R}}_e)_{tt}, \\
\Lambda_{21}^{\mathrm{TE}} &= 4(\bar{\bar{R}}_m)_{zz}.
\end{aligned}
\tag{2.21}
$$

Further, as $(\bar{\bar{\mu}}'_r)_{zz} \to 1$, $(\bar{\bar{R}}_m)_{zz} \to \infty$ and the resultant sheet boundary condition in (2.20) is characterized by continuous tangential electric field components. In the following FE–BI implementation, it is pertinent to define the function $\mathbf{R}_{\mathrm{TE}}$ that maps the electric field in the middle of the sheet, $\mathbf{E}_t = \frac{1}{2}(\mathbf{E}_t^{\mathrm{I}} + \mathbf{E}_t^{\mathrm{II}})$, to the magnetic field, $\Delta \tilde{\mathbf{H}}_t = (\tilde{\mathbf{H}}_t^{\mathrm{I}} - \tilde{\mathbf{H}}_t^{\mathrm{II}})$, so that on Γ_{res}

$$
\hat{n} \times \Delta \tilde{\mathbf{H}}_t = \mathbf{R}_{\mathrm{TE}}(\mathbf{E}_t).
\tag{2.22}
$$

Additionally, for the TM polarization we have

$$
\begin{bmatrix} \gamma_\times \mathbf{E}^{\mathrm{II}} \\ \gamma_t \tilde{\mathbf{H}}^{\mathrm{II}} \end{bmatrix} = \frac{1}{\Delta^{\mathrm{TM}}} \begin{bmatrix} \Lambda_{11}^{\mathrm{TM}} & \Lambda_{12}^{\mathrm{TM}} \\ \Lambda_{21}^{\mathrm{TM}} & \Lambda_{22}^{\mathrm{TM}} \end{bmatrix} \begin{bmatrix} \gamma_\times \mathbf{E}^{\mathrm{I}} \\ \gamma_t \tilde{\mathbf{H}}^{\mathrm{I}} \end{bmatrix},
\tag{2.23}
$$

where $\Delta^{\mathrm{TM}} = [4(\bar{\bar{R}}_e)_{zz}(\bar{\bar{R}}_m)_{tt} - 1]$ and

$$
\begin{aligned}
\Lambda_{11}^{\mathrm{TM}} &= \Lambda_{22}^{\mathrm{TM}} = [4(\bar{\bar{R}}_e)_{zz}(\bar{\bar{R}}_m)_{tt} + 1], \\
\Lambda_{12}^{\mathrm{TM}} &= 4(\bar{\bar{R}}_e)_{zz}, \\
\Lambda_{21}^{\mathrm{TM}} &= 4(\bar{\bar{R}}_m)_{tt}.
\end{aligned}
\tag{2.24}
$$

Again, note that as $(\bar{\bar{\varepsilon}}'_r)_{zz} \to 1$, $(\bar{\bar{R}}_e)_{zz} \to \infty$, and the resultant sheet boundary condition in (2.23) is characterized by continuous tangential magnetic field components. Similar to the expression in (2.22), we can construct function $\mathbf{R}_{\mathrm{TM}}$ on Γ_{res} so that

$$
\hat{n} \times \Delta \tilde{\mathbf{E}}_t = -\mathbf{R}_{\mathrm{TM}}(\mathbf{H}_t).
\tag{2.25}
$$

Finally, we note here that a cross coupling term can be added to (2.19) to produce a more precise representation of the boundary condition for very thin layered media [74].

2.2   SURFACE EQUIVALENCE AND BOUNDARY INTEGRAL EQUATIONS

In the process of truncating the solution domain by Γ_{BI}, we shall see (in the next section) that we must introduce additional unknown quantities to represent the tangential magnetic field, $\tilde{\mathbf{J}}_s$ from (2.9), and the tangential electric field, $\mathbf{M}_s$ from (2.10). As a result, we must derive additional equations to solve for these additional unknowns. To do so, we invoke surface equivalence [28], allowing us to decouple the field solution in the open region outside Γ_{BI} from that inside the bounded region Ω (see Fig. 2.1). Specifically, if both the tangential electric and magnetic fields (i.e., $\mathbf{M}_s$ and $\tilde{\mathbf{J}}_s$) are known everywhere along Γ_{BI}, it can then be shown that the fields radiated/scattered by these equivalent sources are given by

$$\mathbf{E}^{\text{scat}}(\boldsymbol{\rho}) = \Phi(\mathbf{M}_s) - \Theta(\tilde{\mathbf{J}}_s) \tag{2.26}$$

and

$$\tilde{\mathbf{H}}^{\text{scat}}(\boldsymbol{\rho}) = -\Phi(\tilde{\mathbf{J}}_s) - \Theta(\mathbf{M}_s), \tag{2.27}$$

where the integral operators Φ and Θ represent mappings from $\mathcal{H}^{-1/2}(\text{Div}; \partial\Omega)$ to $\mathcal{H}(\text{Curl}; \Omega)$ and have the form

$$\Phi(\mathbf{X}) = \oint_{\Gamma_{BI}} \nabla' G_{2D}(\boldsymbol{\rho}, \boldsymbol{\rho}') \times \mathbf{X}(\boldsymbol{\rho}') \, d\rho', \tag{2.28}$$

$$\Theta(\mathbf{X}) = jk_0 \oint_{\Gamma_{BI}} G_{2D}(\boldsymbol{\rho}, \boldsymbol{\rho}') \mathbf{X}(\boldsymbol{\rho}') \, d\rho' + \frac{j}{k_0} \nabla \oint_{\Gamma_{BI}} G_{2D}(\boldsymbol{\rho}, \boldsymbol{\rho}') \nabla' \cdot \mathbf{X}(\boldsymbol{\rho}') \, d\rho', \tag{2.29}$$

with the 2D Green's function given by $G_{2D}(\boldsymbol{\rho}, \boldsymbol{\rho}') = \frac{j}{4} H_0^{(2)}(k_0 |\boldsymbol{\rho} - \boldsymbol{\rho}'|)$. Here $H_0^{(2)}$ is the zeroth-order Hankel function of the second kind.

Noting that the total field everywhere outside Γ_{BI} must be the sum of the scattered and the incident field, we can state that $\mathbf{M}_s = -\gamma_\times(\mathbf{E}^{\text{inc}} + \mathbf{E}^{\text{scat}})$ and $\tilde{\mathbf{J}}_s = \gamma_\times(\tilde{\mathbf{H}}^{\text{inc}} + \tilde{\mathbf{H}}^{\text{scat}})$ on Γ_{BI}. Substituting in (2.26) and (2.27) gives the boundary integral equations (BIEs)

$$\frac{1}{2}\gamma_\times \mathbf{M}_s - \gamma_t \Phi(\mathbf{M}_s) + \gamma_t \Theta(\tilde{\mathbf{J}}_s) = \gamma_t \mathbf{E}^{\text{inc}}, \tag{2.30}$$

$$\frac{1}{2}\tilde{\mathbf{J}}_s + \gamma_\times \Phi(\tilde{\mathbf{J}}_s) + \gamma_\times \Theta(\mathbf{M}_s) = \gamma_\times \tilde{\mathbf{H}}^{\text{inc}}. \tag{2.31}$$

We note that by forcing the observation point to lie on Γ_{BI}, the integral operator in (2.28) is now a principal value integral excluding the observation point $\boldsymbol{\rho}$. In the literature, (2.30) is most commonly referred to as the electric field integral equation (EFIE) with (2.31) being the magnetic field integral equation (MFIE). We also remark that (2.30) lies in the space $\mathcal{H}^{-1/2}(\text{Curl}; \partial\Omega)$ whereas (2.31) lies in the space $\mathcal{H}^{-1/2}(\text{Div}; \partial\Omega)$.

2.3   VARIATIONAL FORMULATION

Considering TE polarization, the numerical solution of the BVP posed in (2.5) with boundary conditions (2.9), (2.12)–(2.14), (2.17), and (2.22) begins by considering the weak (variational) form of (2.5). Specifically, we seek the solution $\mathbf{E}_t$ everywhere within Ω (see Fig. 2.1) belonging to the vector space [16, 43, 75]

$$\mathcal{W}_E = \{\mathbf{x} \in \mathcal{H}(\text{Curl}; \Omega) \ : \ \gamma_t \mathbf{x}|_{\Gamma_{\text{pec}}} = 0\}. \tag{2.32}$$

Upon realizing the PDE operator in (2.5), $A = \nabla_t \times \frac{1}{\mu_z} \nabla_t \times \ - \ k_0^2 \bar{\bar{\varepsilon}}_r \cdot$ represents a mapping from $\mathcal{H}(\text{Curl}; \Omega)$ to $\mathcal{H}(\text{Div}; \Omega)$ with the corresponding variational form given by

$$\int_\Omega \mathbf{w} \cdot \nabla_t \times \frac{1}{\mu_z} \nabla_t \times \mathbf{E}_t \, d\Omega - k_0^2 \int_\Omega \mathbf{w} \cdot \bar{\bar{\varepsilon}}_r \cdot \mathbf{E}_t \, d\Omega = 0 \qquad \forall \mathbf{w} \in \mathcal{W}_E, \tag{2.33}$$

where we have used the $(\mathcal{L}^2(\Omega))^2$ scalar product. We point out that due to the elliptic nature of the wave equation in (2.5), the sesquilinear form defined in (2.33) is not coercive. However, the uniqueness of the solution in (2.33) follows after a careful analysis [14].

We now perform integration by parts to the first integral kernel in (2.33) using the identity $\mathbf{A} \cdot \nabla \times \mathbf{B} = \nabla \times \mathbf{A} \cdot \mathbf{B} - \nabla \cdot (\mathbf{A} \times \mathbf{B})$ yielding

$$\int_\Omega \nabla_t \times \mathbf{w} \cdot \frac{1}{\mu_z} \nabla_t \times \mathbf{E}_t \, d\Omega - k_0^2 \int_\Omega \mathbf{w} \cdot \bar{\bar{\varepsilon}}_r \cdot \mathbf{E}_t \, d\Omega - \int_\Omega \nabla_t \cdot [\mathbf{w} \times \frac{1}{\mu_z}(\nabla_t \times \mathbf{E}_t)] \, d\Omega = 0. \tag{2.34}$$

After applying the divergence theorem to the last integral in (2.34), we then get

$$\int_\Omega \nabla_t \times \mathbf{w} \cdot \frac{1}{\mu_z} \nabla_t \times \mathbf{E}_t \, d\Omega - k_0^2 \int_\Omega \mathbf{w} \cdot \bar{\bar{\varepsilon}}_r \cdot \mathbf{E}_t \, d\Omega - \int_{\Gamma_{\text{BI}}, \Gamma_{\text{imp}}, \Gamma_{\text{res}}} \mathbf{w} \cdot [\hat{n} \times \frac{1}{\mu_z}(\nabla_t \times \mathbf{E}_t)] \, d\Gamma = 0. \tag{2.35}$$

To solve (2.35), it is necessary to replace the $\nabla \times \mathbf{E}_t$ under the boundary integral with an explicit representation in terms of $\mathbf{E}_t$. The importance of this form is that it concurrently enforces the wave equation and the boundary conditions of the domain in a single mathematical statement. To solve (2.35), we now introduce the boundary conditions satisfied by the fields over Γ_{BI}, Γ_{imp}, and Γ_{res}, derived in Section 2.1.3. This yields

$$\int_\Omega \nabla_t \times \mathbf{w} \cdot \frac{1}{\mu_z} \nabla_t \times \mathbf{E}_t \, d\Omega - k_0^2 \int_\Omega \mathbf{w} \cdot \bar{\bar{\varepsilon}}_r \cdot \mathbf{E}_t \, d\Omega + jk_0 \oint_{\Gamma_{\text{BI}}} \mathbf{w} \cdot \tilde{\mathbf{J}}_s \, d\Gamma$$

$$+ jk_0 \int_{\Gamma_{\text{imp}}} \mathbf{w} \cdot \frac{1}{\eta_{tt}} \gamma_t \mathbf{E}_t \, d\Gamma + jk_0 \int_{\Gamma_{\text{res}}} \mathbf{w} \cdot \mathbf{R}_{\text{TE}}(\mathbf{E}_t) \, d\Gamma = 0, \tag{2.36}$$

where an additional unknown $\tilde{\mathbf{J}}_s$ now being explicitly shown under the integral over Γ_{BI}. Thus, one more equation is needed to supplement (2.36). For this, we introduce the EFIE as in (2.30).

That is to say, we seek $\tilde{\mathbf{J}}_s \in \mathcal{V}_J$, where

$$\mathcal{V}_J = \{\mathbf{x} \in \mathcal{H}^{-1/2}(\mathrm{Div}; \Gamma_{\mathrm{BI}}) \; : \; \mathbf{x}|_{\Gamma_{\mathrm{BI}} \cap \Gamma_{\mathrm{pmc}}} = 0\}, \tag{2.37}$$

such that

$$\frac{1}{2}\oint_{\Gamma_{\mathrm{BI}}} \mathbf{v} \cdot \mathbf{E}_t \, d\Gamma + \oint_{\Gamma_{\mathrm{BI}}} \mathbf{v} \cdot \Phi(\hat{n} \times \mathbf{E}_t) \, d\Gamma$$
$$+ \oint_{\Gamma_{\mathrm{BI}}} \mathbf{v} \cdot \Theta(\tilde{\mathbf{J}}_s) \, d\Gamma = \oint_{\Gamma_{\mathrm{BI}}} \mathbf{v} \cdot \mathbf{E}^{\mathrm{inc}} \, d\Gamma \qquad \forall \mathbf{v} \in \mathcal{V}_J. \tag{2.38}$$

In typical FE–BI formulations [37], the set of equations (2.36) and (2.38) are combined to generate a coupled hybrid matrix system (we neglect interior resonance issues [63–71] at this point). However, this system is not symmetric (i.e., the matrix resulting from the discretization of the system is not symmetric). To obtain a symmetric system, in a manner similar to that in [22], one way is to consider the variational form of the MFIE as in (2.31). This is actually an integral equation of the second kind and, thus, its variational form is similar to that in (1.17) where we seek $\tilde{\mathbf{J}}_s \in \mathcal{V}_J$ such that

$$\frac{1}{2}\oint_{\Gamma_{\mathrm{BI}}} \mathbf{w} \cdot \tilde{\mathbf{J}}_s \, d\Gamma = -\oint_{\Gamma_{\mathrm{BI}}} \mathbf{w} \cdot \hat{n} \times \Phi(\tilde{\mathbf{J}}_s) \, d\Gamma + \oint_{\Gamma_{\mathrm{BI}}} \mathbf{w} \cdot \hat{n} \times \Theta(\hat{n} \times \mathbf{E}_t) \, d\Gamma$$
$$+ \oint_{\Gamma_{\mathrm{BI}}} \mathbf{w} \cdot \hat{n} \times \tilde{\mathbf{H}}^{\mathrm{inc}} \, d\Gamma \qquad \forall \mathbf{w}|_{\Gamma_{\mathrm{BI}}} \in \mathcal{W}_E. \tag{2.39}$$

Unlike (2.38), Eq. (2.39) has testing function space dual to $\mathcal{V}_J$ as required by Theorem 3 in Chapter 1. Combining (2.36) with (2.39) we get

$$\frac{1}{jk_0}\int_{\Omega} \nabla_t \times \mathbf{w} \cdot \frac{1}{\mu_z} \nabla_t \times \mathbf{E}_t \, d\Omega + jk_0 \int_{\Omega} \mathbf{w} \cdot \bar{\bar{\varepsilon}}_r \cdot \mathbf{E}_t \, d\Omega - \oint_{\Gamma_{\mathrm{BI}}} (\hat{n} \times \mathbf{w}) \cdot \Theta(\hat{n} \times \mathbf{E}_t) \, d\Gamma$$
$$+ \frac{1}{2}\oint_{\Gamma_{\mathrm{BI}}} \mathbf{w} \cdot \tilde{\mathbf{J}}_s \, d\Gamma + \oint_{\Gamma_{\mathrm{BI}}} (\hat{n} \times \mathbf{w}) \cdot \Phi(\tilde{\mathbf{J}}_s) \, d\Gamma + \int_{\Gamma_{\mathrm{imp}}} \mathbf{w} \cdot \frac{1}{\eta_{tt}} \gamma_t \mathbf{E}_t \, d\Gamma \tag{2.40}$$
$$+ \int_{\Gamma_{\mathrm{res}}} \mathbf{w} \cdot \mathbf{R}_{\mathrm{TE}}(\mathbf{E}_t) \, d\Gamma = \oint_{\Gamma_{\mathrm{BI}}} (\hat{n} \times \mathbf{w}) \cdot \tilde{\mathbf{H}}^{\mathrm{inc}} \, d\Gamma.$$

In the next section, we show that the system of equations (2.40) and (2.38) can be combined under certain circumstances to produce a symmetric matrix upon discretization. This approach was used recently [22] and allows for a more robust implementation of the standard FE–BI [37].

In the FE–BI system (2.40) and (2.38), we assumed that $\Gamma_{\mathrm{imp}} \cap \Gamma_{\mathrm{BI}} = \{\emptyset\}$ and $\Gamma_{\mathrm{res}} \cap \Gamma_{\mathrm{BI}} = \{\emptyset\}$. That is, no impedance or sheet boundary condition was defined on the boundary

Γ_{BI}. Of course, if $\Gamma_{\text{imp}} \cap \Gamma_{\text{BI}} \neq \{\emptyset\}$, then the fields inside Ω should be eliminated and the BI equation in (2.38) alone should be used to calculate the field solution, using $\hat{n} \times \mathbf{E}_t = \eta_{tt}\gamma_{\times}\tilde{\mathbf{J}}_s$ from (2.15). Also, if $\Gamma_{\text{res}} \cap \Gamma_{\text{BI}} \neq \{\emptyset\}$, then the only change to the FE–BI system is to assume that $\mathbf{E}_t|_{\Gamma_{\text{BI}}} = \mathbf{E}_t^{\text{I}}$ and $\mathbf{E}_t|_\Omega = \mathbf{E}_t^{\text{II}}$ on the surface Γ_{res}. That is, for all integral kernels over Γ_{BI} we have that $\mathbf{E}_t = \mathbf{E}_t^{\text{I}}$; and for all integral kernels over Ω we set $\mathbf{E}_t = \mathbf{E}_t^{\text{II}}$.

It appears that we have yet to explicitly enforce the PMC boundary condition within Ω (note that: the material and PEC boundary conditions were explicitly enforced by the choice of solution space $\mathcal{W}_E$). However, this boundary condition is intrinsically satisfied in a weak sense by the FE–BI system. All that is done in the implementation of a PMC boundary condition is not to explicitly enforce tangential field continuity across the PMC boundaries (see Chapter 3 of [37]).

We end this section by simply stating the symmetric variational formulation for TM polarization. That is, we seek $\tilde{\mathbf{H}}_t \in \mathcal{W}_H$ and $\mathbf{M}_s \in \mathcal{V}_M$, such that

$$
\frac{1}{jk_0} \int_\Omega \nabla_t \times \mathbf{w} \cdot \frac{1}{\varepsilon_z} \nabla_t \times \tilde{\mathbf{H}}_t \, d\Omega + jk_0 \int_\Omega \mathbf{w} \cdot \bar{\mu}_r \cdot \tilde{\mathbf{H}}_t \, d\Omega - \oint_{\Gamma_{\text{BI}}} (\hat{n} \times \mathbf{w}) \cdot \Theta(\hat{n} \times \tilde{\mathbf{H}}_t) \, d\Gamma
$$
$$
+ \frac{1}{2} \oint_{\Gamma_{\text{BI}}} \mathbf{w} \cdot \mathbf{M}_s \, d\Gamma + \oint_{\Gamma_{\text{BI}}} (\hat{n} \times \mathbf{w}) \cdot \Phi(\mathbf{M}_s) \, d\Gamma + \int_{\Gamma_{\text{imp}}} \mathbf{w} \cdot \eta_{zz}\gamma_t \tilde{\mathbf{H}}_t \, d\Gamma
$$
$$
+ \int_{\Gamma_{\text{res}}} \mathbf{w} \cdot \mathbf{R}_{\text{TM}}(\tilde{\mathbf{H}}_t) \, d\Gamma = -\oint_{\Gamma_{\text{BI}}} (\hat{n} \times \mathbf{w}) \cdot \mathbf{E}^{\text{inc}} \, d\Gamma.
$$

$$(2.41)$$

for all $\mathbf{w} \in \mathcal{W}_H$, and

$$
\frac{1}{2} \oint_{\Gamma_{\text{BI}}} \mathbf{v} \cdot \tilde{\mathbf{H}}_t \, d\Gamma + \oint_{\Gamma_{\text{BI}}} \mathbf{v} \cdot \Phi(\hat{n} \times \tilde{\mathbf{H}}_t) \, d\Gamma
$$
$$
+ \oint_{\Gamma_{\text{BI}}} \mathbf{v} \cdot \Theta(\mathbf{M}_s) \, d\Gamma = \oint_{\Gamma_{\text{BI}}} \mathbf{v} \cdot \mathbf{H}^{\text{inc}} \, d\Gamma \qquad \forall \mathbf{v} \in \mathcal{V}_M,
$$

$$(2.42)$$

where

$$
\mathcal{W}_H = \{\mathbf{x} \in \mathcal{H}(\text{Curl}; \Omega) \; : \; \gamma_t \mathbf{x}|_{\Gamma_{\text{pmc}}} = 0\},
$$

$$(2.43)$$

and

$$
\mathcal{V}_M = \{\mathbf{x} \in \mathcal{H}^{-1/2}(\text{Div}; \Gamma_{\text{BI}}) \; : \; \mathbf{x}|_{\Gamma_{\text{BI}} \cap \Gamma_{\text{pec}}} = 0\}.
$$

$$(2.44)$$

2.4   DISCRETIZATION

In this section, we proceed to discretize the variational formulations so that they can be solved numerically. As described in Section 1.2, we partition the bounded domain Ω into a discretized domain Ω_h using the curl conforming parametric finite elements defined in (1.76). For the boundary we must employ the divergence conforming elements defined in (1.74) (see Figs. 1.7 and 1.10). Thus, if the domain Ω_h consists of N_{FE} finite elements and N_{BI} boundary elements, then for TE polarization, the unknowns $\mathbf{E}_t$ and $\tilde{\mathbf{J}}_s$ can be expanded as

$$\mathbf{E}_t(\mathbf{r}) = \sum_{i=1}^{N_{\mathrm{FE}}} \sum_{j=1}^{4} \chi_j^i \, \mathbf{w}_j^i(\mathbf{r}(u, v)), \tag{2.45}$$

and

$$\tilde{\mathbf{J}}_s(\mathbf{r}) = \sum_{i=1}^{N_{\mathrm{BI}}} \sum_{j=1}^{2} \upsilon_j^i \, \mathbf{v}_j^i(\mathbf{r}(u)), \tag{2.46}$$

where the vector functions $\mathbf{w}_j^i$ are congruent to those defined in (1.76) and $\mathbf{v}_j^i$ are congruent to those defined in (1.74). For $\mathbf{E}_t \in \mathcal{W}_E$ and $\tilde{\mathbf{J}}_s \in \mathcal{V}_J$, we need to first eliminate those unknowns χ_j^i and υ_j^i that correspond to PEC and PMC edges, respectively. We must also pair all remaining unknowns in both $\mathbf{E}_t$ and $\tilde{\mathbf{J}}_s$ expansions that share edges within Ω_h (any unpaired current unknown is also eliminated). These pairings ensure that $\mathbf{E}_t$ is curl conforming and $\tilde{\mathbf{J}}_s$ is divergence conforming. Once these steps are completed, the expansions in both (2.45) and (2.46) can be more succinctly expressed as

$$\begin{aligned} \mathbf{E}_t(\mathbf{r}) = \sum_{k=1}^{N_E} \chi_k^E \, \mathbf{w}_k^E(\mathbf{r}) + \sum_{k=1}^{N_I} \chi_k^I \, \mathbf{w}_k^I(\mathbf{r}) \\ + \sum_{k=1}^{N_R} \chi_k^R \, \mathbf{w}_k^R(\mathbf{r}) + \sum_{k=1}^{N_M} \chi_k^M \, \mathbf{w}_k^M(\mathbf{r}), \end{aligned} \tag{2.47}$$

and

$$\tilde{\mathbf{J}}_s(\mathbf{r}) = \sum_{k=1}^{N_J} \upsilon_k \, \mathbf{v}_k(\mathbf{r}), \tag{2.48}$$

where N^E is equal to the number of non-PEC edges in Ω_h not associated with the boundaries Γ_{BI}, Γ_{imp} or Γ_{res}; N^I and N^R are equal to the number of element edges on Γ_{imp} and Γ_{res}, respectively; N^M is the number of non-PEC edges on Γ_{BI}; N^J is equal to the number of non-PMC edges on Γ_{BI}. The expansion functions defined in (2.47) and (2.48), consisting of linear combinations of at most two *local* functions $\mathbf{w}_j^i$ and $\mathbf{v}_j^i$, are often termed *global* expansion basis functions.

To construct the system matrix (see Section 1.1.2), we substitute (2.47) and (2.48) into the variational equations (2.40) and (2.38) to obtain the matrix system

$$
\begin{bmatrix}
\bar{Z}^{EE} & \bar{Z}^{ER} & \bar{Z}^{EI} & \bar{Z}^{EM} & \bar{0} \\
\bar{Z}^{RE} & \bar{Z}^{RR} & \bar{Z}^{RI} & \bar{Z}^{RM} & \bar{0} \\
\bar{Z}^{IE} & \bar{Z}^{IR} & \bar{Z}^{II} & \bar{Z}^{IM} & \bar{0} \\
\bar{Z}^{ME} & \bar{Z}^{MR} & \bar{Z}^{MI} & \bar{Z}^{MM} & \bar{Z}^{MJ} \\
\bar{0} & \bar{0} & \bar{0} & \bar{Z}^{JM} & \bar{Z}^{JJ}
\end{bmatrix}
\begin{bmatrix}
\mathbf{E}^{E} \\
\mathbf{E}^{I} \\
\mathbf{E}^{R} \\
\mathbf{E}^{M} \\
\mathbf{J}
\end{bmatrix}
=
\begin{bmatrix}
\mathbf{0} \\
\mathbf{0} \\
\mathbf{0} \\
\mathbf{b} \\
\mathbf{f}
\end{bmatrix},
\tag{2.49}
$$

where the unknown vectors represent columns, namely $\mathbf{E}^{\#} = \{\chi_k^{\#}\}_{k=1}^{N^{\#}}$ with $\# = \{E, I, R, M\}$ and $\mathbf{J} = \{v_k\}_{k=1}^{N^J}$. If we define the matrix $\bar{A}^{XY}$ as

$$
\bar{A}_{ij}^{XY} = \frac{1}{jk_0} \int_{\Omega_j^Y} \frac{1}{\mu_z} \nabla_t \times \mathbf{w}_i^X \cdot \nabla_t \times \mathbf{w}_j^Y \, d\Omega + jk_0 \int_{\Omega_j^Y} \mathbf{w}_i^X \cdot \bar{\bar{\varepsilon}}_r \mathbf{w}_j^Y \, d\Omega,
\tag{2.50}
$$

then the system matrices in (2.49) take the explicit forms:

$$
\begin{aligned}
& \bar{Z}^{EE} = \bar{A}^{EE}, \quad \bar{Z}^{ER} = \bar{A}^{ER}, \quad \bar{Z}^{EI} = \bar{A}^{EI}, \quad \bar{Z}^{EM} = \bar{A}^{EM}, \\
& \bar{Z}^{RE} = \bar{A}^{RE}, \quad \bar{Z}^{RI} = \bar{A}^{RI}, \quad \bar{Z}^{RM} = \bar{A}^{RM}, \\
& \bar{Z}^{IE} = \bar{A}^{IE}, \quad \bar{Z}^{IR} = \bar{A}^{IR}, \quad \bar{Z}^{IM} = \bar{A}^{IM}, \\
& \bar{Z}^{ME} = \bar{A}^{ME}, \quad \bar{Z}^{MR} = \bar{A}^{MR}, \quad \bar{Z}^{MI} = \bar{A}^{MI}, \\
& \bar{Z}_{ij}^{RR} = \bar{A}_{ij}^{RR} + \int_{\Gamma_{\text{res}}^j} \mathbf{w}_i^R \cdot \mathbf{R}_{\text{TE}}(\mathbf{w}_j^R) \, d\Gamma, \\
& \bar{Z}_{ij}^{II} = \bar{A}_{ij}^{II} + \int_{\Gamma_{\text{imp}}^j} \frac{1}{\eta_{tt}} \mathbf{w}_i^I \cdot \mathbf{w}_j^I \, d\Gamma, \\
& \bar{Z}_{ij}^{MM} = \bar{A}_{ij}^{MM} - \int_{\Gamma_{\text{BI}}^i} (\hat{n} \times \mathbf{w}_i^M) \cdot \Theta(\hat{n} \times \mathbf{w}_j^M) \, d\Gamma, \\
& \bar{Z}_{ij}^{MJ} = \frac{1}{2} \int_{\Gamma_{\text{BI}}^j} \mathbf{w}_i^M \cdot \mathbf{v}_j \, d\Gamma + \int_{\Gamma_{\text{BI}}^i} (\hat{n} \times \mathbf{w}_i^M) \cdot \Phi(\mathbf{v}_j) \, d\Gamma, \\
& \bar{Z}_{ij}^{JM} = \frac{1}{2} \int_{\Gamma_{\text{BI}}^j} \mathbf{v}_i \cdot \mathbf{w}_j^M \, d\Gamma + \int_{\Gamma_{\text{BI}}^i} \mathbf{v}_i \cdot \Phi(\hat{n} \times \mathbf{w}_j^M) \, d\Gamma, \\
& \bar{Z}_{ij}^{JJ} = \int_{\Gamma_{\text{BI}}^i} \mathbf{v}_i \cdot \Theta(\mathbf{v}_j) \, d\Gamma.
\end{aligned}
\tag{2.51}
$$

Further, the excitation vectors are given by

$$
\mathbf{b}_i = \int_{\Gamma_{\text{BI}}^i} (\hat{n} \times \mathbf{w}_i^M) \cdot \tilde{\mathbf{H}}^{\text{inc}} \, d\Gamma,
$$

$$
\mathbf{f}_i = \int_{\Gamma_{\text{BI}}^i} \mathbf{v}_i \cdot \mathbf{E}^{\text{inc}} \, d\Gamma.
\tag{2.52}
$$

It can be shown that as long as $\bar{\bar{\varepsilon}}_r$ is reciprocal, the resulting FE–BI system matrix in (2.49) is symmetric.

We note that the integral kernels described in (2.50) and (2.51) can be evaluated numerically using any Gaussian quadrature scheme such as the Gauss–Legendre quadrature [76]. However, when evaluating the integral kernels involving the singular operators Φ and Θ one must use a singularity annihilation scheme such as the Duffy transform [77,78].

2.5   EXAMPLE DISCRETIZATION

To clarify the process of obtaining the global expansions in (2.47) and (2.48) from the local expansions in (2.45) and (2.46), let us consider a rather simple mesh depicted in Figs. 2.3 and 2.4. We can conclude from the mesh in Fig. 2.3 that there are 12 element edges within Ω_b and 12 element edges on Γ_{BI} (only material boundary conditions are specified throughout this structure since our focus is on demonstrating the process). Note that the numbering scheme in Fig. 2.3 depicts the relationship between the local and global FE unknowns. Explicitly, we make use of the unknown assembly in Tables 2.1 and 2.2 to demonstrate how the pairing process is accomplished for the FE unknowns. For the boundary unknowns, we make use of the numbering scheme presented in Fig. 2.4 to construct the corresponding assembly in Table 2.3. It is important to note that these pairings ensure both tangential field continuity across element edges as well as normal current continuity across boundary element edges (recall that an "edge" for a boundary element in 2D is actually a node joining two boundary elements together).

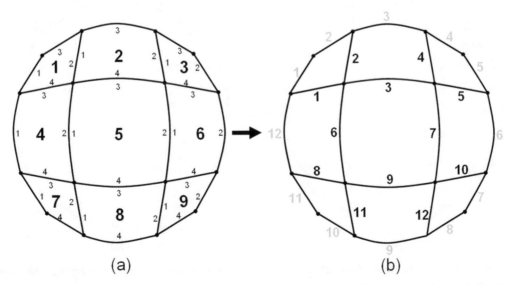

FIGURE 2.3: Sample mesh demonstrating FE unknown assembly: (a) local numbering, (b) global numbering

TABLE 2.1: Nonboundary FE Unknown Assembly

LOCAL TO GLOBAL UNKNOWN RELATIONSHIPS		
$\chi_1^E = \chi_4^1 = \chi_3^4$	$\chi_5^E = \chi_4^3 = \chi_3^6$	$\chi_9^E = \chi_4^5 = \chi_3^8$
$\chi_2^E = \chi_2^1 = \chi_1^2$	$\chi_6^E = \chi_2^4 = \chi_1^5$	$\chi_{10}^E = \chi_4^6 = \chi_3^9$
$\chi_3^E = \chi_4^2 = \chi_3^5$	$\chi_7^E = \chi_2^5 = \chi_1^6$	$\chi_{11}^E = \chi_2^7 = \chi_1^8$
$\chi_4^E = \chi_2^2 = \chi_1^3$	$\chi_8^E = \chi_4^4 = \chi_3^7$	$\chi_{12}^E = \chi_2^8 = \chi_1^9$

TABLE 2.2: Boundary FE Unknown Assembly

LOCAL TO GLOBAL UNKNOWN RELATIONSHIPS		
$\chi_1^M = \chi_1^1$	$\chi_5^M = \chi_2^3$	$\chi_9^M = \chi_4^8$
$\chi_2^M = \chi_3^1$	$\chi_6^M = \chi_2^6$	$\chi_{10}^M = \chi_4^7$
$\chi_3^M = \chi_3^2$	$\chi_7^M = \chi_2^9$	$\chi_{11}^M = \chi_1^7$
$\chi_4^M = \chi_3^3$	$\chi_8^M = \chi_4^9$	$\chi_{12}^M = \chi_1^4$

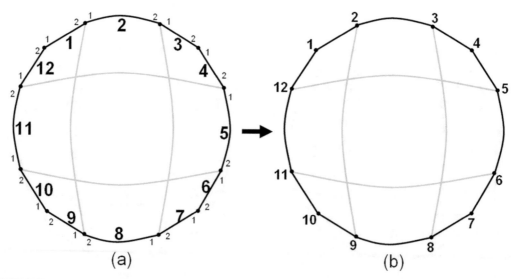

FIGURE 2.4: Sample mesh demonstrating BI unknown assembly: (a) local numbering, (b) global numbering

TABLE 2.3: BI Unknown Assembly
LOCAL TO GLOBAL UNKNOWN RELATIONSHIPS

$$v_1 = v_1^1 = v_2^{12} \qquad v_5 = v_1^5 = v_2^4 \qquad v_9 = v_1^9 = v_2^8$$

$$v_2 = v_1^2 = v_2^1 \qquad v_6 = v_1^6 = v_2^5 \qquad v_{10} = v_1^{10} = v_2^9$$

$$v_3 = v_1^3 = v_2^2 \qquad v_7 = v_1^7 = v_2^6 \qquad v_{11} = v_1^{11} = v_2^{10}$$

$$v_4 = v_1^4 = v_2^3 \qquad v_8 = v_1^8 = v_2^7 \qquad v_{12} = v_1^{12} = v_2^{11}$$

Once the unknown pairings are defined through the assembly tables, we proceed to define the global expansion functions. For example, the geometry depicted in Fig. 2.3 with unknown pairings given in Table 2.1 lead to the global FE expansion function $\mathbf{w}_1^E$ given by $\mathbf{w}_1^E = \mathbf{w}_4^1 + \mathbf{w}_3^4$. Similarly, from Table 2.3 the global BI expansion function $\mathbf{v}_1$ becomes $\mathbf{v}_1 = \mathbf{v}_1^1 + \mathbf{v}_2^{12}$. We remark that all the unknown pairings described in this example, the local expansion functions at the edges, were assumed to be equivalent. However, in some cases, the direction of the two paired basis functions may be opposite. For example, if for $\mathbf{w}_4^1$ and $\mathbf{w}_3^4$ we have that $\mathbf{w}_4^1 = -\mathbf{w}_3^4$, then the unknown pairing becomes $\chi_1^E = \chi_4^1 = -\chi_3^4$ and the global expansion function must be revised to read $\mathbf{w}_1^E = \mathbf{w}_4^1 - \mathbf{w}_3^4$. This can also be accomplished by assigning a "sense" (or orientation) to each global edge and each local edge. The orientation of the local edge are, of course, determined by the definitions of the local basis functions in (2.45) and (2.46). The sign correction then falls out when there is a mismatch in the "sense" of the global edge and the local edge.

Once the unknown assembly tables are known for a given numerical simulation, the table can then be used to construct the system matrices described in (2.49). To observe this, consider again the example mesh in Fig. 2.3. Typically, in numerical implementations, one first constructs element-to-element interaction matrices, and then proceeds to place the resulting values into the proper place within the proper system matrix. Essentially, an element-to-element interaction matrix represents the interaction of one element's local expansion functions with those of another. For those FE matrices that do not involve the Φ and/or Θ operators, the element-to-element interaction matrices are only nonzero when the testing and source elements are equivalent (hence they are usually called "element" matrices). However, this is not the case when considering the element-to-element interaction for the boundary integral matrices where all matrix terms are nonzero and dense. These facts manifest themselves in the well-known property of having sparse FE matrices and dense BI matrices.

In Fig. 2.5 we demonstrate how the 4×4 element-to-element interaction matrices are used in conjunction with the unknown assembly tables to place the resulting matrix values into their proper place within the FE–BI system matrices. Specifically, Fig. 2.5 shows how the $\bar{Z}_{11}^{EE}$

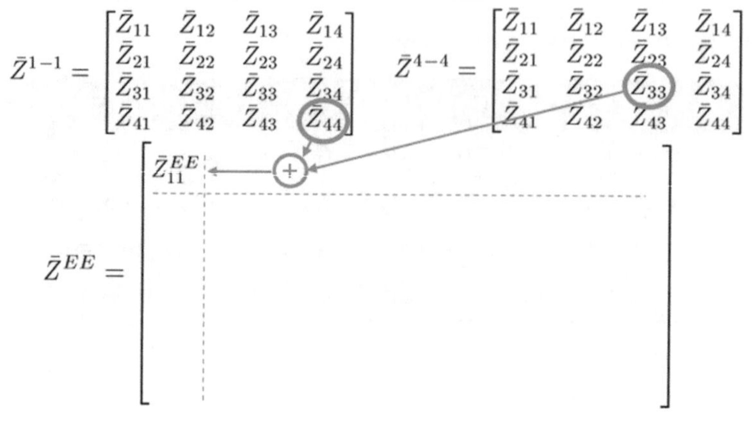

$$\bar{Z}^{1-1} = \begin{bmatrix} \bar{Z}_{11} & \bar{Z}_{12} & \bar{Z}_{13} & \bar{Z}_{14} \\ \bar{Z}_{21} & \bar{Z}_{22} & \bar{Z}_{23} & \bar{Z}_{24} \\ \bar{Z}_{31} & \bar{Z}_{32} & \bar{Z}_{33} & \bar{Z}_{34} \\ \bar{Z}_{41} & \bar{Z}_{42} & \bar{Z}_{43} & \boxed{\bar{Z}_{44}} \end{bmatrix} \qquad \bar{Z}^{4-4} = \begin{bmatrix} \bar{Z}_{11} & \bar{Z}_{12} & \bar{Z}_{13} & \bar{Z}_{14} \\ \bar{Z}_{21} & \bar{Z}_{22} & \bar{Z}_{23} & \bar{Z}_{24} \\ \bar{Z}_{31} & \bar{Z}_{32} & \boxed{\bar{Z}_{33}} & \bar{Z}_{34} \\ \bar{Z}_{41} & \bar{Z}_{42} & \bar{Z}_{43} & \bar{Z}_{44} \end{bmatrix}$$

$$\bar{Z}^{EE} = \begin{bmatrix} \bar{Z}^{EE}_{11} \leftarrow \boxed{+} \\ \\ \\ \\ \end{bmatrix}$$

FIGURE 2.5: Matrix assembly demonstration

submatrix is constructed as a sum of both $\bar{Z}_{44}$ from the 1–1 interaction matrix and the $\bar{Z}_{33}$ from the 4–4 interaction matrix. This results from the fact that $\chi_1^E = \chi_4^1 = \chi_3^4$ from Table 2.1.

At this point the reader is ready to implement their own FE–BI program. In the next section, we discuss applications of the presented formulation to model scattering from arbitrary 2D cylinders.

2.6   2D SCATTERING APPLICATIONS

In the previous sections, we describe the development of a hybrid 2D FE–BI formulation. The discretization scheme discussed in Section 2.4 allows for accurate implementation of the formulation on personal computers. In this section, we provide a few examples of the formulation's utility for scattering applications. Many of the results validation examples are taken from the report by Syed and Volakis [79].

Typically, when calculating the scattering from an object, we are interested in measuring the object's radar cross section (RCS), or echo width (for 2D objects). This is given by

$$\sigma_{\text{RCS}}^{2D}(\phi) = \lim_{\rho \to \infty} 2\pi\rho \frac{|\mathbf{E}^{\text{scat}}(\phi)|^2}{|\mathbf{E}^{\text{inc}}|^2} = \lim_{\rho \to \infty} 2\pi\rho \frac{|\tilde{\mathbf{H}}^{\text{scat}}(\phi)|^2}{|\tilde{\mathbf{H}}^{\text{inc}}|^2}. \qquad (2.53)$$

For our purposes we will only consider plane wave excitations. That is, for TE polarizations, we have

$$\tilde{\mathbf{H}}^{\text{inc}} = e^{-jk_0(\hat{k}^{\text{inc}} \cdot \rho)}\hat{z}$$
$$\mathbf{E}^{\text{inc}} = e^{-jk_0(\hat{k}^{\text{inc}} \cdot \rho)}(\hat{z} \times \hat{k}^{\text{inc}}), \qquad (2.54)$$

and for TM polarizations, we have

$$\mathbf{E}^{\text{inc}} = e^{-jk_0(\hat{k}^{\text{inc}} \cdot \rho)}\hat{z}$$
$$\tilde{\mathbf{H}}^{\text{inc}} = e^{-jk_0(\hat{k}^{\text{inc}} \cdot \rho)}(\hat{k}^{\text{inc}} \times \hat{z}), \qquad (2.55)$$

where the incident wave direction $\hat{k}^{\text{inc}}$ is given by $\hat{k}^{\text{inc}} = -(\hat{x}\cos\phi_{\text{inc}} + \hat{y}\sin\phi_{\text{inc}})$. Additionally, once the tangential fields are known everywhere on the boundary Γ_{BI}, the scattered electric or magnetic far-field can be calculated using the far-field approximation to either (2.26) or (2.27), respectively. For the sake of brevity, we simply state here the far-field approximations resulting from Φ and Θ operators:

$$\Phi^{\text{FF}}(\mathbf{X}) = e^{j\pi/4}\sqrt{\frac{k_0}{8\pi}}\frac{e^{-jk_0\rho}}{\sqrt{\rho}}\hat{\rho}\times\oint_{\Gamma_{\text{BI}}}\mathbf{X}(\boldsymbol{\rho}')e^{jk_0\boldsymbol{\rho}'\cdot\hat{\rho}}\,d\rho', \qquad (2.56)$$

$$\Theta^{\text{FF}}(\mathbf{X}) = -e^{j\pi/4}\sqrt{\frac{k_0}{8\pi}}\frac{e^{-jk_0\rho}}{\sqrt{\rho}}\hat{\rho}\times\hat{\rho}\times\oint_{\Gamma_{\text{BI}}}\mathbf{X}(\boldsymbol{\rho}')e^{jk_0\boldsymbol{\rho}'\cdot\hat{\rho}}\,d\rho'. \qquad (2.57)$$

Here, the observation direction $\hat{\rho}$ is given by $\hat{\rho} = \hat{x}\cos\phi + \hat{y}\sin\phi$.

The first two scattering objects considered in this section are the simple PEC square and circular cylinders shown in the insets of Figs. 2.6 and 2.7. As seen, both the square cylinder's width and the circular cylinder's radius have length 1λ. Though, it is possible to simulate these structures using only a BI formulation, where Γ_{BI} would be chosen conformal to the surface of the square, we instead choose Γ_{BI} to enclose the cylinders such that a $\lambda/20$ thick free-space layer is formed over their surfaces. In so doing, we can confirm the FE–BI's capability to handle the PEC boundary condition within Ω over both flat and curved surfaces. It can be seen in

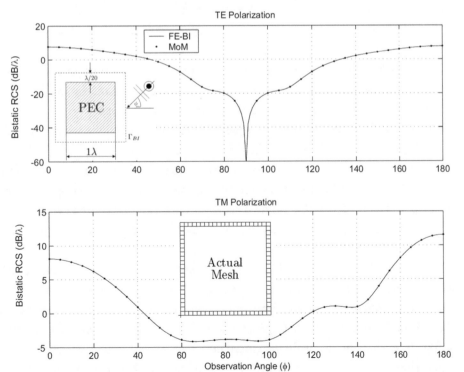

FIGURE 2.6: TE/TM scattering from a PEC square cylinder (MoM data from Syed and Volakis [79])

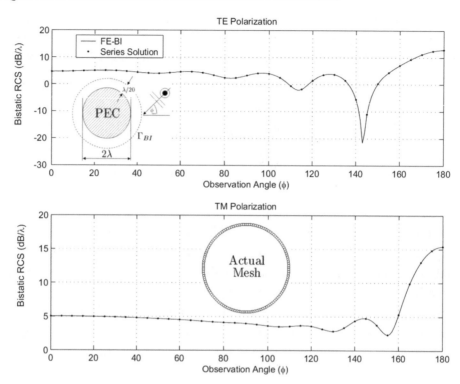

FIGURE 2.7: TE/TM scattering from a PEC circular cylinder (MoM data from Syed and Volakis [79])

Fig. 2.6 that the FE–BI solution is accurate with respect to the method of moments (MoM) solution obtained using only a BI implementation. In addition, the results in Fig. 2.7 confirm the accuracy of the FE–BI when compared to eigenfunction solutions.

We next consider a slightly more complex scattering object where we add a 0.15λ thick material coating having $\varepsilon_r = 3 - j3$ to the PEC circular cylinder considered in Fig. 2.7 (see inset in Fig. 2.8). This coated circular cylinder was simulated using the FE–BI formulation and the results are compared to those obtained using analytic solutions in Fig. 2.8. Because we again force Γ_{BI} to lie on a $\lambda/20$ thick free-space layer enclosing the coated cylinder, this result primarily confirms the FE–BI's capability to model fields inside regions consisting of more than one type of material.

In the next application of the 2D FE–BI, we choose to confirm the formulation's ability to model impedance surfaces. Note that in Section 2.1.3 we mentioned how these boundary surfaces are most often used to model imperfect conductors or PEC surfaces having very thin material coatings. In either case, an impedance surface should only be defined over all or part of a closed surface where fields inside the closed boundary are assumed zero. We see in Fig. 2.9 the accurate simulation of an impedance cylinder of radius 1λ having a relative impedance $\eta = 2 - j2$. The FE–BI is compared to the standard moment method (MoM in the figure).

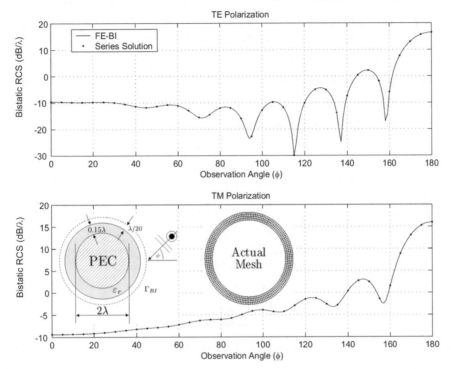

FIGURE 2.8: TE/TM scattering from a coated PEC circular cylinder (MoM data from Syed and Volakis [79])

To exploit the FE–BI's capability to model more complex structures we now consider the scattering from a square cylinder of edge length 1λ having thin material coatings on two sides of the cylinder (see inset in Fig. 2.10). In addition, we consider the case where a 0.15λ thick coating of $\varepsilon_r = 3 - j3$ is applied to the square cylinder (see inset in Fig. 2.11). For the impedance boundaries, we use the same impedance value used in the previous example, namely, $\eta = 2 - j2$. The results in Figs. 2.10 and 2.11 confirm the flexibility of the hybrid FE–BI formulation. One can imagine that the most sophisticated low observable coatings do indeed require computational tools that can model very complex structures. Obviously, the results presented here demonstrate that FE–BI is well suited for such applications.

We continue by considering the scattering from objects consisting of anisotropic materials. In Fig. 2.12, we model the scattering from a composite square cylinder of edge length $\lambda/2$ having material parameters

$$\bar{\bar{\varepsilon}}_r = \bar{\bar{\mu}}_r = \begin{bmatrix} \frac{\pi}{2} - j & 0 & 0 \\ 0 & \pi - j\frac{1}{2} & 0 \\ 0 & 0 & \frac{\pi}{2} \end{bmatrix}. \tag{2.58}$$

Because we chose $\bar{\bar{\varepsilon}}_r = \bar{\bar{\mu}}_r$, we should expect both TE and TM solutions in Fig. 2.12 to be equivalent.

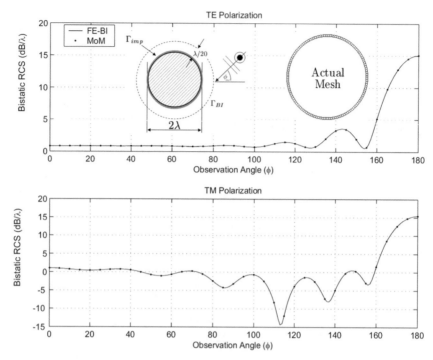

FIGURE 2.9: TE/TM scattering from an impedance circular cylinder (MoM data from Syed and Volakis [79])

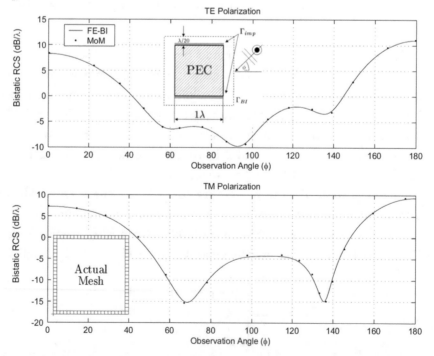

FIGURE 2.10: TE/TM scattering from an impedance/PEC square cylinder (MoM data from Syed and Volakis [79])

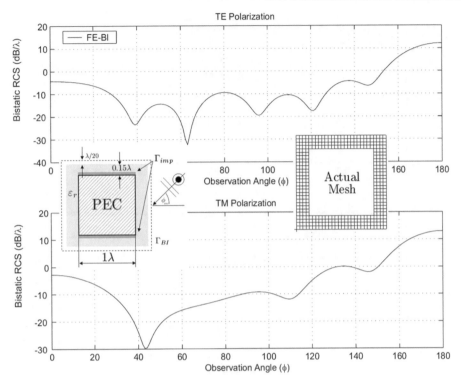

FIGURE 2.11: TE/TM scattering from a coated impedance/PEC square cylinder

The proposed symmetric FE–BI formulation is not impervious to interior resonances [72]. That is, at some frequencies—corresponding to the resonances of the cavity created by the boundary—the field solution is not unique. This causes the condition number of the system matrix to drastically increase and can cause the solution to be incorrect. A good explanation of the interior resonance problem and some typical solutions can be found in [80]. To see the effects of interior resonances on the condition number of the system matrix, we refer to Fig. 2.13 where we show the calculated condition number over a given frequency range for the system matrix corresponding to an air-filled cylinder having 1m radius. As seen, the condition number spikes at those frequencies equal to the cavity resonances [28]. In Fig. 2.13, we also include the effects of diagonal scaling on the condition number, where we see similar results to those obtained in [72].

Finally, we consider TM scattering from the same PEC circular cylinder used in [71]. This cylinder is enclosed by a circular boundary Γ_{BI} of radius $a_0 = 1.01a_c$ for the FE–BI solution, where a_c is the radius of the actual PEC circular cylinder. In Fig. 2.14 [71], we see that when a typical FE–BI formulation is employed, the monostatic RCS data are corrupted at frequencies equal to the interior resonances of the cylinder created by the Γ_{BI} boundary. However, as seen in Fig. 2.15, with diagonal scaling alone, the symmetric FE–BI formulation performs well

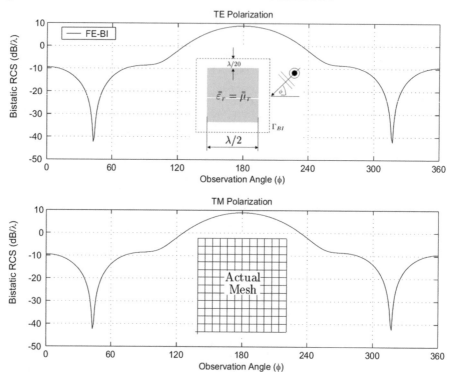

FIGURE 2.12: TE/TM scattering from a composite anisotropic square cylinder

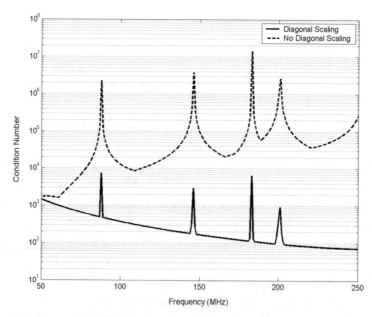

FIGURE 2.13: Condition number of the system matrix for an air-filled cylinder with radius 1 m

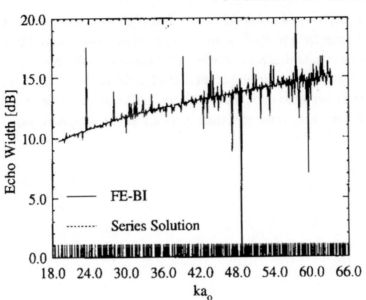

FIGURE 2.14: Backscatter echo width versus ka_0 for a PEC circular cylinder—the lines over the horizontal axis correspond to the eigenvalues of a circular waveguide (after Collins et al., © IEEE, 1992 [71])

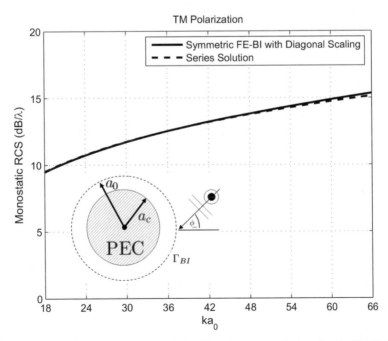

FIGURE 2.15: Comparison of the symmetric FE–BI to the series solution for the PEC circular cylinder in Fig. 2.14.

(for this simulation, we used a biconjugate gradient (BiCG) iterative solver with the iterative convergence tolerance set to 10^{-4}). This result suggests that the symmetric FE–BI formulation is rather robust. However, further consideration is needed to solve the internal resonance problem without destroying the symmetry properties of the matrix. In the next chapter, we introduce a 3D FE–BI formulation based upon the combined field integral equation (CFIE) that does not contain internal resonances at the expense of having asymmetric system matrices.

CHAPTER 3

Three-Dimensional Hybrid FE–BI: Formulation and Applications

Open-domain electromagnetic scattering and radiation by complex inhomogeneous structures in three dimensions is of great importance. With their application to radar cross section (RCS) prediction, antenna analysis, microwave circuit design, electromagnetic coupling/interference characterization, and inverse scattering, electromagnetic simulations of 3D open-domain problems for composite structures have received much attention in the past two decades. In Chapter 2, we introduced a hybrid method based upon a coupling of the finite element method (FEM) with the boundary element method (BEM). There, the boundary elements were introduced to exactly satisfy the radiation condition on the truncated finite element domain.

Some of the first engineering applications of this method to 3D structures can be found in [47, 60, 81, 82] for scattering/RCS applications and [83] for antenna radiation problems. For a review of the early developments of this method in three dimensions, we refer the reader to [84–88]. A drawback of the method is the necessity to solve a partially dense matrix system resulting from the boundary element formulation. As such, early applications of this method were often restricted to planar boundaries so that iterative solution techniques such as CG-FFT could be utilized to reduce memory and CPU requirements down to *O(N)* [47, 48]. However, with the development of fast iterative solvers including the fast multipole method (FMM) [89, 90], the FE–BI method has been ubiquitously adapted to solve many practical scattering and antenna problems [91–93].

In this chapter, we describe the FE–BI formulation for 3D scattering and radiation problems. As seen in the previous chapter (2D FE–BI description), use of the BEM on closed domains can produce spurious solutions at frequencies corresponding to the natural modes of the closed boundary. To circumvent this problem, we employ the combined field integral equation (CFIE) for the boundary element formulation [68] (though there are a myriad of other methods to solve the interior resonance problem [63–67, 69–71]).

We begin by describing the PDE and boundary conditions that make up the boundary value problem (BVP) to be solved. To end up with a system of equations, we next develop the integral equation that will be included within the FEM system. We then present the variational statement for the hybrid FE–BI formulation allowing for a numerical implementation through a discretization consisting of the finite elements discussed in Section 1.3. In the last section, we verify the versatility of the method with a presentation of scattering and antenna applications.

3.1   THE BOUNDARY VALUE PROBLEM

Let us assume a material region Ω characterized by $\bar{\bar{\varepsilon}}_r(\mathbf{r})$ and $\bar{\bar{\mu}}_r(\mathbf{r})$ with an impressed current source $\mathbf{J}^{\mathrm{imp}}(\mathbf{r})$ (see Fig. 3.1) within Ω. The electromagnetic fields within the domain obey the vector wave equation given by (as described on page 5 of [37])

$$\nabla \times \bar{\bar{\mu}}_r^{-1}(\mathbf{r}) \cdot \nabla \times \mathbf{E}(\mathbf{r}) - k_0^2 \bar{\bar{\varepsilon}}_r(\mathbf{r}) \cdot \mathbf{E}(\mathbf{r}) = \tilde{\mathbf{J}}^{\mathrm{imp}}(\mathbf{r}). \tag{3.1}$$

Here, k_0 is the free-space wavenumber and the relative anisotropic permittivity and permeability dyads are given by

$$\bar{\bar{\varepsilon}}_r(\mathbf{r}) = \begin{bmatrix} \varepsilon_{xx} & \varepsilon_{xy} & \varepsilon_{xz} \\ \varepsilon_{yx} & \varepsilon_{yy} & \varepsilon_{yz} \\ \varepsilon_{zx} & \varepsilon_{zy} & \varepsilon_{zz} \end{bmatrix} \quad \text{and} \quad \bar{\bar{\mu}}_r(\mathbf{r}) = \begin{bmatrix} \mu_{xx} & \mu_{xy} & \mu_{xz} \\ \mu_{yx} & \mu_{yy} & \mu_{yz} \\ \mu_{zx} & \mu_{zy} & \mu_{zz} \end{bmatrix}. \tag{3.2}$$

From our functional analysis discussion in Chapter 1, it can easily be seen that we must have $\mathbf{E} \in \mathcal{H}(\mathrm{Curl}; \Omega)$ and $\tilde{\mathbf{J}}^{\mathrm{imp}} \in \mathcal{H}(\mathrm{Div}; \Omega)$.

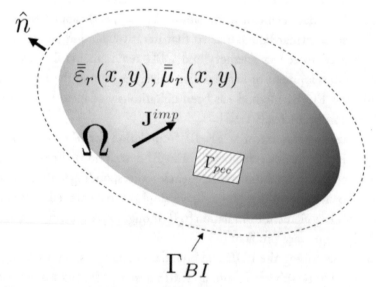

FIGURE 3.1: Three-dimensional FE–BI domain

In general, one can consider all the boundary conditions discussed in Section 2.1.3. However, for the sake of brevity, we only consider here the PEC, PMC, and radiation boundary conditions. Essentially, we seek the solution of the electric field $\mathbf{E}$ for all $\mathbf{r} \in \Omega$, such that $\mathbf{E}$ satisfies (3.1) everywhere subject to (see Eq. (1.29) for definition of $\gamma_\times$)

$$\gamma_t \mathbf{E} = 0 \qquad \mathbf{r} \in \Gamma_{\text{PEC}}, \tag{3.3}$$

$$\gamma_\times \bar{\bar{\mu}}_r^{-1} \cdot \nabla \times \mathbf{E} = 0 \qquad \mathbf{r} \in \Gamma_{\text{PMC}}, \tag{3.4}$$

$$\gamma_\times \bar{\bar{\mu}}_r^{-1} \cdot \nabla \times \mathbf{E} = -jk_0 \tilde{\mathbf{J}}_s, \ \mathbf{r} \in \Gamma_{\text{BI}}. \tag{3.5}$$

We note that the equivalent surface current $\tilde{\mathbf{J}}_s = Z_0 \gamma_\times \mathbf{H} \in \mathcal{H}^{-1/2}(\text{Div}; \partial\Omega)$ has been introduced on the domain boundary Γ_{BI} in (3.5). We will see that the application of this boundary condition on Γ_{BI} introduces an additional electromagnetic unknown into the formulation. This implies a need for an additional equation to obtain a solvable numerical system.

3.2   BOUNDARY INTEGRAL EQUATIONS

The surface equivalence principle [28] states that if the tangential electric and magnetic fields ($\mathbf{M}_s$ and $\tilde{\mathbf{J}}_s$) are known everywhere on some closed boundary (see Fig. 3.2), then the fields everywhere in a homogeneous medium outside this boundary can be uniquely determined from the integral representations

$$\mathbf{E}^{\text{scat}}(\mathbf{r}) = \Phi(\mathbf{M}_s) - \Theta(\tilde{\mathbf{J}}_s) \tag{3.6}$$

and

$$\tilde{\mathbf{H}}^{\text{scat}}(\mathbf{r}) = -\Phi(\tilde{\mathbf{J}}_s) - \Theta(\mathbf{M}_s). \tag{3.7}$$

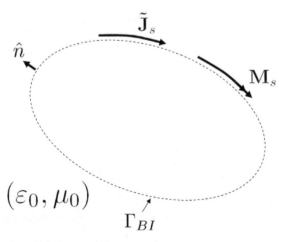

FIGURE 3.2: Boundary integral diagram

Here the integral operators Φ and Θ represent mappings from $\mathcal{H}^{-1/2}(\mathrm{Div}; \partial\Omega)$ to $\mathcal{H}(\mathrm{Curl}; \Omega)$ and have the form

$$\Phi(\mathbf{X}) = \oint_{\Gamma_{\mathrm{BI}}} \nabla' G_{3\mathrm{D}}(\mathbf{r}, \mathbf{r}') \times \mathbf{X}(\mathbf{r}') \, dr', \tag{3.8}$$

$$\Theta(\mathbf{X}) = jk_0 \oint_{\Gamma_{\mathrm{BI}}} G_{3\mathrm{D}}(\mathbf{r}, \mathbf{r}')\mathbf{X}(\mathbf{r}') \, dr' + \frac{j}{k_0} \nabla \oint_{\Gamma_{\mathrm{BI}}} G_{3\mathrm{D}}(\mathbf{r}, \mathbf{r}')\nabla' \cdot \mathbf{X}(\mathbf{r}') \, dr', \tag{3.9}$$

with the 3D Green's function given by $G_{3\mathrm{D}}(\mathbf{r}, \mathbf{r}') = \frac{e^{-jk_0|\mathbf{r}-\mathbf{r}'|}}{4\pi|\mathbf{r}-\mathbf{r}'|}$ and $\nabla' G_{3\mathrm{D}} = -\nabla G_{3\mathrm{D}}$.

For radiation problems, where only an impressed current source is considered, the scattered fields $\mathbf{E}^{\mathrm{scat}}$ and $\tilde{\mathbf{H}}^{\mathrm{scat}}$ become the radiated fields $\mathbf{E}^{\mathrm{rad}}$ and $\tilde{\mathbf{H}}^{\mathrm{rad}}$. Nevertheless, the total field everywhere outside Γ_{BI} is the sum of the scattered/radiated field and the incident field (if it exists). Thus, on the boundary surface Γ_{BI}, $\mathbf{M}_s = -\gamma_\times(\mathbf{E}^{\mathrm{inc}} + \mathbf{E}^{\mathrm{scat}})$ and $\tilde{\mathbf{J}}_s = \gamma_\times(\tilde{\mathbf{H}}^{\mathrm{inc}} + \tilde{\mathbf{H}}^{\mathrm{scat}})$. Upon combining these with (3.6) and (3.7), we can derive the following boundary integral equations (BIEs).

$$\frac{1}{2}\gamma_\times\mathbf{M}_s - \gamma_t\Phi(\mathbf{M}_s) + \gamma_t\Theta(\tilde{\mathbf{J}}_s) = \gamma_t\mathbf{E}^{\mathrm{inc}}, \tag{3.10}$$

$$\frac{1}{2}\tilde{\mathbf{J}}_s + \gamma_\times\Phi(\tilde{\mathbf{J}}_s) + \gamma_\times\Theta(\mathbf{M}_s) = \gamma_\times\tilde{\mathbf{H}}^{\mathrm{inc}}. \tag{3.11}$$

We note that by forcing the observation point to lie on Γ_{BI}, the integral operator in (3.8) is now a principal value integral excluding the observation point $\mathbf{r}$ (see page 230 of [37]). In the literature, (3.10) is most commonly referred to as the electric field integral equation (EFIE) whereas (3.11) is correspondingly the magnetic field integral equation (MFIE). Although either of these BIEs can be used to complete the FE–BI formulation, they both suffer from interior resonance problems. One approach to overcome the interior resonance issues is to linearly combine the two equations to form the so-called combined field integral equation (CFIE) given by

$$\alpha\left[\frac{1}{2}\gamma_\times\mathbf{M}_s - \gamma_t\Phi(\mathbf{M}_s) + \gamma_t\Theta(\tilde{\mathbf{J}}_s)\right] + (1-\alpha)\left[\frac{1}{2}\tilde{\mathbf{J}}_s + \gamma_\times\Phi(\tilde{\mathbf{J}}_s) + \gamma_\times\Theta(\mathbf{M}_s)\right]$$
$$= \alpha\gamma_t\mathbf{E}^{\mathrm{inc}} + (1-\alpha)\gamma_\times\tilde{\mathbf{H}}^{\mathrm{inc}}. \tag{3.12}$$

Here the real constant $\alpha \in [0, 1]$ is used as a weighting parameter and based on experience α should be chosen to favor the EFIE equation (i.e., $\alpha > 0.5$).

3.3   THE FE–BI VARIATIONAL STATEMENT

We are now ready to express the boundary value problem in its variational form. To do so, we must choose the proper functional space in which we seek the solution of the electric field $\mathbf{E}$. Specifically, if we define the vector space $\mathcal{W}_E$ as in Section 2.3, namely

$$\mathcal{W}_E = \{\mathbf{x} \in \mathcal{H}(\mathrm{Curl}; \Omega) : \gamma_t\mathbf{x}|_{\Gamma_{\mathrm{PEC}}} = 0\}, \tag{3.13}$$

then the variational statement requires us to seek $\mathbf{E} \in \mathcal{W}_E$ such that

$$\int_\Omega \mathbf{w} \cdot \nabla \times \bar{\bar{\mu}}_r^{-1} \cdot \nabla \times \mathbf{E} \, d\Omega - k_0^2 \int_\Omega \mathbf{w} \cdot \bar{\bar{\varepsilon}}_r \cdot \mathbf{E} \, d\Omega = jk_0 \int_\Omega \mathbf{w} \cdot \tilde{\mathbf{J}}^{\mathrm{imp}} \, d\Omega \qquad \forall \mathbf{w} \in \mathcal{W}_E. \quad (3.14)$$

As is typical of all variational forms of second-order PDEs, we proceed to apply integration by parts to the left integrand of (3.14) yielding

$$\int_\Omega \nabla \times \mathbf{w} \cdot \bar{\bar{\mu}}_r^{-1} \cdot \nabla \times \mathbf{E} \, d\Omega - k_0^2 \int_\Omega \mathbf{w} \cdot \bar{\bar{\varepsilon}}_r \cdot \mathbf{E} \, d\Omega + jk_0 \oint_{\Gamma_{\mathrm{BI}}} \mathbf{w} \cdot \tilde{\mathbf{J}}_s \, d\Gamma = jk_0 \int_\Omega \mathbf{w} \cdot \tilde{\mathbf{J}}^{\mathrm{imp}} \, d\Omega,$$

$$(3.15)$$

where we have used the boundary condition (3.5) in the third integrand of (3.15). Because of the unknown $\tilde{\mathbf{J}}_s$ (in addition to $\mathbf{E}$), on the boundary Γ_{BI}, an additional equation is needed. The introduction of the EFIE, MFIE, or CFIE to relate $\tilde{\mathbf{J}}_s$ and $\mathbf{E}$ on Γ_{BI} leads to the traditional FE–BI formulation. Using the CFIE (to avoid interior resonances), we have the BI equation

$$\alpha \frac{1}{2} \int_{\Gamma_{\mathrm{BI}}} \mathbf{v} \cdot \mathbf{E} \, d\Gamma + \alpha \int_{\Gamma_{\mathrm{BI}}} \mathbf{v} \cdot \Phi(\hat{n} \times \mathbf{E}) \, d\Gamma + (1 - \alpha) \int_{\Gamma_{\mathrm{BI}}} (\hat{n} \times \mathbf{v}) \cdot \Theta(\hat{n} \times \mathbf{E}) \, d\Gamma$$

$$+ (1 - \alpha) \frac{1}{2} \int_{\Gamma_{\mathrm{BI}}} \mathbf{v} \cdot \tilde{\mathbf{J}}_s \, d\Gamma - (1 - \alpha) \int_{\Gamma_{\mathrm{BI}}} (\hat{n} \times \mathbf{v}) \cdot \Phi(\tilde{\mathbf{J}}_s) \, d\Gamma + \alpha \int_{\Gamma_{\mathrm{BI}}} \mathbf{v} \cdot \Theta(\tilde{\mathbf{J}}_s) \, d\Gamma \quad (3.16)$$

$$= \alpha \int_{\Gamma_{\mathrm{BI}}} \mathbf{v} \cdot \mathbf{E}^{\mathrm{inc}} \, d\Gamma - (1 - \alpha) \int_{\Gamma_{\mathrm{BI}}} (\hat{n} \times \mathbf{v}) \cdot \tilde{\mathbf{H}}^{\mathrm{inc}} \, d\Gamma \qquad \forall \mathbf{v} \in \mathcal{V}_J.$$

such that $\tilde{\mathbf{J}}_s \in \mathcal{V}_J$ where

$$\mathcal{V}_J = \{ \mathbf{x} \in \mathcal{H}^{-1/2}(\mathrm{Div}; \Gamma_{\mathrm{BI}}) \ : \ \mathbf{x}|_{\Gamma_{\mathrm{BI}} \cap \Gamma_{\mathrm{PMC}}} = 0 \}. \quad (3.17)$$

The combined set of (3.15) and (3.12) render the FE–BI formulation for solving $\mathbf{E}$ within $\mathcal{W}_E$ and $\tilde{\mathbf{J}}_s$ on $\mathcal{V}_J$. We remark that this system of equations is not symmetric like the formulation derived in Chapter 2. However, this system is free of the interior resonance problem depicted in Fig. 2.13.

3.4   DISCRETIZATION

We now proceed to discretize the FE–BI variational Eqs. (3.15) and (3.16) for numerical solution. As in Section 2.4, we partition the bounded domain Ω into a discretized domain Ω_h consisting of the curl conforming 3D parametric finite element defined in (1.77). The boundary Ω_h, is modeld by the divergence conforming boundary elements defined in (1.75) (see Figs. 1.9 and 1.11). Thus, if the domain Ω_h consists of N_{FE} finite elements and N_{BI} boundary elements, the unknowns $\mathbf{E}$ and $\tilde{\mathbf{J}}_s$ can be expanded as

$$\mathbf{E}(\mathbf{r}) = \sum_{i=1}^{N_{\mathrm{FE}}} \sum_{j=1}^{12} \chi_j^i \, \mathbf{w}_j^i (\mathbf{r}(u, v, w)), \quad (3.18)$$

and

$$\tilde{\mathbf{J}}_s(\mathbf{r}) = \sum_{i=1}^{N_{\mathrm{BI}}} \sum_{j=1}^{4} \upsilon_j^i \, \mathbf{v}_j^i (\mathbf{r}(u, v)), \tag{3.19}$$

where the vector functions $\mathbf{w}_j^i$ are congruent to those defined in (1.77) and $\mathbf{v}_j^i$ are congruent to those defined in (1.75). For $\mathbf{E} \in \mathcal{W}_E$ and $\tilde{\mathbf{J}}_s \in \mathcal{V}_J$, we need to first eliminate those unknowns, χ_j^i and υ_j^i, that correspond to PEC and PMC edges, respectively. We must also pair all remaining unknowns in both the $\mathbf{E}$ and $\tilde{\mathbf{J}}_s$ expansions that correspond to the same edges within Ω_h, a process referred to as the "assembly." Also, any unpaired current unknowns are eliminated. These pairings ensure that $\mathbf{E}$ is curl conforming and $\tilde{\mathbf{J}}_s$ is divergence conforming. Once these steps are completed, the expansions in both (3.18) and (3.19) can be more succinctly expressed as

$$\mathbf{E}(\mathbf{r}) = \sum_{k=1}^{N_E} \chi_k^E \, \mathbf{w}_k^E(\mathbf{r}) + \sum_{k=1}^{N_M} \chi_k^M \, \mathbf{w}_k^M(\mathbf{r}), \tag{3.20}$$

and

$$\tilde{\mathbf{J}}_s(\mathbf{r}) = \sum_{k=1}^{N_J} \upsilon_k \, \mathbf{v}_k(\mathbf{r}), \tag{3.21}$$

where N^E is equal to the number of non-PEC edges in Ω_h, not associated with the boundaries Γ_{BI}; N^M is the number of non-PEC edges on Γ_{BI}; N^J is equal to the number of non-PMC edges on Γ_{BI}. We note that the expansion functions defined in (3.20) and (3.21), consisting of a linear combination of at most two *local* functions $\mathbf{w}_j^i$ and $\mathbf{v}_j^i$, are often termed *global* expansion functions. Fig. 3.3 represents one such vector basis function defined on the support of two adjacent hexahedral elements.

To construct the system matrix (see Section 1.1.2), we substitute (3.20) and (3.21) into the variational Eqs. (3.15) and (3.16) to obtain the matrix system

$$\begin{bmatrix} \bar{Z}^{EE} & \bar{Z}^{EM} & \bar{0} \\ \bar{Z}^{ME} & \bar{Z}^{MM} & \bar{Z}^{MJ} \\ \bar{0} & \bar{Z}^{JM} & \bar{Z}^{JJ} \end{bmatrix} \begin{bmatrix} \mathbf{E}^E \\ \mathbf{E}^B \\ \mathbf{J} \end{bmatrix} = \begin{bmatrix} \mathbf{f}^E \\ \mathbf{f_M} \\ \mathbf{g} \end{bmatrix}, \tag{3.22}$$

where the unknown vectors represent columns, namely, $\mathbf{E}^\# = \{\chi_k^\#\}_{k=1}^{N^\#}$ with $\# = \{E, M\}$ and $\mathbf{J} = \{\upsilon_k\}_{k=1}^{N^J}$. If we define the matrix $\bar{A}^{XY}$ as

$$\bar{A}_{ij}^{XY} = \int_{\Omega_j^Y} \nabla \times \mathbf{w}_i^X \cdot \bar{\mu}_r^{-1} \cdot \nabla \times \mathbf{w}_j^Y \, \mathrm{d}\Omega - k_0^2 \int_{\Omega_j^Y} \mathbf{w}_i^X \cdot \bar{\varepsilon}_r \cdot \mathbf{w}_j^Y \, \mathrm{d}\Omega, \tag{3.23}$$

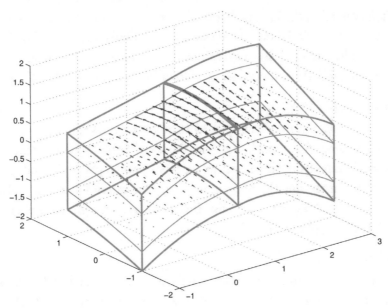

FIGURE 3.3: Illustration of the variation in second-order electric field basis functions used in the context of 3D FE–BI. The arrows indicate the field direction and their length is proportional to their strength

the system matrices in (3.22) take the explicit forms

$$\bar{\bar{Z}}^{EE} = \bar{\bar{A}}^{EE}, \qquad \bar{\bar{Z}}^{EM} = \bar{\bar{A}}^{EM},$$

$$\bar{\bar{Z}}^{ME} = \bar{\bar{A}}^{ME}, \qquad \bar{\bar{Z}}^{MM} = \bar{\bar{A}}^{MM}, \; \bar{Z}_{ij}^{MJ} = jk_0 \int_{\Gamma_{\mathrm{BI}}^i} \mathbf{w}_i^M \cdot \mathbf{v}_j \, d\Gamma$$

$$\bar{Z}_{ij}^{JM} = \frac{\alpha}{2} \int_{\Gamma_{\mathrm{BI}}^i} \mathbf{v}_i \cdot \mathbf{w}_j^M \, d\Gamma + \alpha \int_{\Gamma_{\mathrm{BI}}^i} \mathbf{v}_i \cdot \Phi\left(\hat{n} \times \mathbf{w}_j^M\right) \, d\Gamma + (1-\alpha) \int_{\Gamma_{\mathrm{BI}}^i} (\hat{n} \times \mathbf{v}_i) \cdot \Theta\left(\hat{n} \times \mathbf{w}_j^M\right) \, d\Gamma$$

$$\bar{Z}_{ij}^{JJ} = (1-\alpha)\frac{1}{2} \int_{\Gamma_{\mathrm{BI}}^i} \mathbf{v}_i \cdot \mathbf{v}_j \, d\Gamma - (1-\alpha) \int_{\Gamma_{\mathrm{BI}}^i} (\hat{n} \times \mathbf{v}_i) \cdot \Phi(\mathbf{v}_j) \, d\Gamma + \alpha \int_{\Gamma_{\mathrm{BI}}^i} \mathbf{v}_j \cdot \Theta(\mathbf{v}_j) \, d\Gamma.$$

$$(3.24)$$

Further, the excitation vector is given by

$$\mathbf{f}_i^{\#} = jk_0 \int_{\Omega_i^{\#}} \mathbf{w}_i^{\#} \cdot \tilde{\mathbf{J}}^{\mathrm{imp}} \, d\Omega$$

$$\mathbf{g}_i = \alpha \int_{\Gamma_{\mathrm{BI}}^i} \mathbf{v}_i \cdot \mathbf{E}^{\mathrm{inc}} \, d\Gamma - (1-\alpha) \int_{\Gamma_{\mathrm{BI}}^i} (\hat{n} \times \mathbf{v}_i) \cdot \tilde{\mathbf{H}}^{\mathrm{inc}} \, d\Gamma.$$

$$(3.25)$$

We note that the integral kernels described in (3.23) and (3.24) can be evaluated numerically using any Gaussian quadrature scheme such as the Gauss–Legendrè quadrature [76].

However, when evaluating the integral kernels involving the singular operators Φ and Θ, one must use a singularity annihilation scheme such as the Duffy transform [77, 78]. For details of the process of filling the FE–BI matrix from the local matrices, we refer the reader to the discussion in Section 2.5. Also, the reader is referred to Chapters 3 and 4 of the book [37].

3.5   APPLICATIONS

In this section, we provide various examples to demonstrate the wide range of problems where the FE–BI is often used. Advances in both the theory of finite element method coupled with the developments in fast algorithms as well as the ever improving capacity and speed of modern computing platforms make the FE–BI method a very accurate and powerful tool in analyzing a vast variety of real-life engineering problems. Below, we present some example radar scattering and antenna radiation applications.

3.5.1   Scattering

The definition of the radar cross section (RCS, referred to as echo area in 3D and echo width in 2D) for two dimensions was presented in Section 2.6. For 3D objects, the RCS is correspondingly defined as

$$\sigma_{\text{RCS}}^{\text{3D}}(\phi, \theta) = \lim_{r \to \infty} 4\pi r^2 \frac{|\mathbf{E}^{\text{scat}}(\phi, \theta)|^2}{|\mathbf{E}^{\text{inc}}|^2} = \lim_{r \to \infty} 4\pi r^2 \frac{|\tilde{\mathbf{H}}^{\text{scat}}(\phi, \theta)|^2}{|\tilde{\mathbf{H}}^{\text{inc}}|^2}. \qquad (3.26)$$

As in Chapter 2, we will only consider plane wave excitations. That is, we consider excitations of the form

$$\begin{aligned} \mathbf{E}^{\text{inc}} &= \hat{E}^{\text{inc}} e^{-jk_0 \hat{k}^{\text{inc}} \cdot \mathbf{r}}, \\ \tilde{\mathbf{H}}^{\text{inc}} &= \hat{k}^{\text{inc}} \times \mathbf{E}^{\text{inc}} \end{aligned} \qquad (3.27)$$

where the incident wave direction $\hat{k}^{\text{inc}}$ is given by

$$\hat{k}^{\text{inc}} = -(\hat{x} \cos \phi_{\text{inc}} \sin \theta_{\text{inc}} + \hat{y} \sin \phi_{\text{inc}} \sin \theta_{\text{inc}} + \hat{z} \cos \theta_{\text{inc}}) \qquad (3.28)$$

and the impinging electric field polarization $\hat{E}^{\text{inc}}$ is given by

$$\begin{bmatrix} \hat{E}_x^{\text{inc}} \\ \hat{E}_y^{\text{inc}} \\ \hat{E}_z^{\text{inc}} \end{bmatrix} = E_\theta^{\text{inc}} \begin{bmatrix} \cos \phi_{\text{inc}} \cos \theta_{\text{inc}} \\ \sin \phi_{\text{inc}} \cos \theta_{\text{inc}} \\ -\sin \theta_{\text{inc}} \end{bmatrix} + E_\phi^{\text{inc}} \begin{bmatrix} -\sin \phi_{\text{inc}} \\ \cos \phi_{\text{inc}} \\ 0 \end{bmatrix}. \qquad (3.29)$$

Typically, one is interested in the RCS for both, θ ($E_\phi^{\text{inc}} = 0$) and ϕ ($E_\theta^{\text{inc}} = 0$) polarized incident waves (in some texts these are referred to as vertical and horizontal polarization, respectively). One may also be interested in the RCS scattering matrix defined by

$$\bar{\bar{\sigma}}_{\text{RCS}}^{\text{3D}} = \begin{bmatrix} \sigma_{\theta,\theta} & \sigma_{\theta,\phi} \\ \sigma_{\phi,\theta} & \sigma_{\phi,\phi} \end{bmatrix}, \qquad (3.30)$$

where

$$\sigma_{X,Y}(\phi, \theta) = \lim_{r \to \infty} 4\pi r^2 \frac{|\mathbf{E}_X^{\text{scat}}(\phi, \theta)|^2}{|\mathbf{E}_Y^{\text{inc}}|^2} \quad \text{for} \quad X, Y \in \{\theta, \phi\}. \tag{3.31}$$

After the tangential fields are known everywhere on the boundary Γ_{BI}, the scattered electric or magnetic far-field can be calculated using the far-field approximation to either (3.6) or (3.7), respectively. For the sake of brevity, we simply state here the far-field approximations resulting from the Φ and Θ operators, namely,

$$\Phi^{\text{FF}}(\mathbf{X}) = \frac{jk_0}{4\pi} \frac{e^{-jk_0 r}}{r} \hat{r} \times \oint_{\Gamma_{\text{BI}}} \mathbf{X}(\mathbf{r}')e^{jk_0\mathbf{r}' \cdot \hat{r}} \, d\mathbf{r}', \tag{3.32}$$

$$\Theta^{\text{FF}}(\mathbf{X}) = -\frac{jk_0}{4\pi} \frac{e^{-jk_0 r}}{r} \hat{r} \times \hat{r} \times \oint_{\Gamma_{\text{BI}}} \mathbf{X}(\mathbf{r}')e^{jk_0\mathbf{r}' \cdot \hat{r}} \, d\mathbf{r}', \tag{3.33}$$

where $\hat{r} = \hat{x} \cos\phi \sin\theta + \hat{y} \sin\phi \sin\theta + \hat{z} \cos\theta$ is the observation direction. We note that the corresponding far-field $\mathbf{E}$ ($\mathbf{E}^{\text{FF}}$) can be obtained by substituting (3.32) and (3.33) into (3.6).

Our first example is that of scattering by a solid lossy dielectric sphere of $\epsilon_r = 1.75 - 0.3j$. This is an electrically small problem having a radius of only $0.2\lambda_0$, where λ_0 is the free-space wavelength at the solution frequency. The FE–BI solution for two different volume tessellations are depicted in Fig. 3.4 along with the reference Mie solution. As observed in the figure, the FE–BI solution converges to the reference Mie series result as the mesh density is increased.

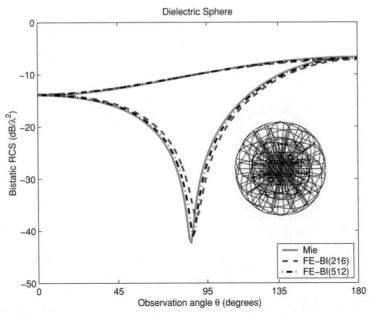

FIGURE 3.4: Bistatic RCS of a dielectric sphere using the FE–BI method

However, the convergence behavior of the BiCG iterative solver [37] used for the denser mesh problem (with 512 hexahedra) was rather poor. Quantitatively, the larger problem results in 1176 volume unknowns and 768 surface unknowns. Aimed at improving this poor convergence behavior, we employed a diagonal preconditioner. This allowed the BiCG solver to converge in 302 iterations (executing 2 matrix–vector products per iteration) to within an error of 10^{-2}.

Similar poor iterative convergence behavior was observed for our second example, that of a dielectric coated PEC sphere. The outer radius of the coated sphere was $0.75\lambda_0$ and the coating itself had a thickness $0.075\lambda_0$ with a lossy $\epsilon_r = 1.75 - 0.3\text{j}$. A total of 384 curvilinear hexahedra were used to model the coating and this problem resulted in a higher number of surface unknowns (1536) and fewer FE unknowns (386). The bistatic RCS results for both polarizations of the incident field are given in Fig. 3.5. For this case, the BiCG solver with diagonal preconditioning took 85 iterations to converge to an error of 10^{-2}. These observations suggest that the iterative convergence behavior of the FE–BI system is severely deteriorated by the presence of large FE domains. Such poor convergence behavior is due to the highly heterogeneous nature of the generated FE–BI system, which is partly sparse due to the FE domain and partly dense due to the BI operator. One remedy is to use better preconditioners [94,95]. However, the specific choice and performance of the preconditioner depends on the geometry under investigation and puts a computational burden on the solver when electrically

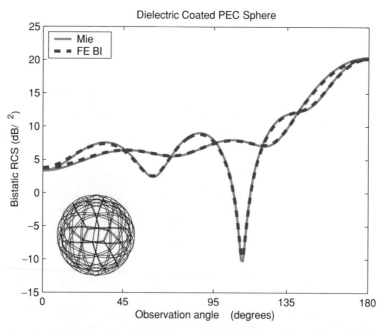

FIGURE 3.5: Bistatic RCS of a coated sphere using the FE–BI method. The sphere was $0.75\lambda_0$ in outer radius with a $0.075\lambda_0$ thick coating having a relative permittivity of $\epsilon_r = 1.75 - 0.3\text{j}$

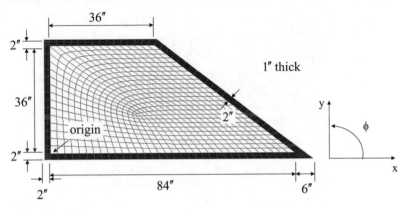

FIGURE 3.6: Trapezoidal test target

large problems are considered. Nevertheless, when a very large number of decoupled FEM domains are considered, as in the case with large finite antenna arrays, efficient block diagonal preconditioning using the self FE–BI matrix of each antenna element has been demonstrated to considerably improve convergence [93] leading to a very successful application of the method.

The last scattering example is a trapezoidal target having a lossy material coating ($\epsilon = 4.5 - 9.0j$) around the edges [96, 97]. It was modeled using 1320 hexahedral finite elements as depicted in Fig. 3.6. The monostatic RCS results for both polarizations are shown in Fig. 3.7. As seen, for this example there is a large dynamic range in the computed RCS data (40 dB). Also, the presence of nonmetallic sections make the geometry modeling challenging. This is clearly an example where the FE–BI is advantageous because of its geometrical adaptability and material generalities.

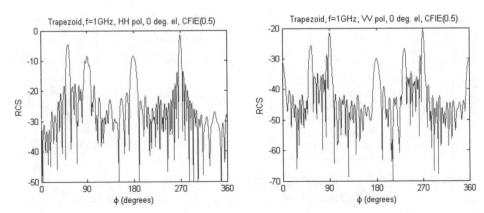

FIGURE 3.7: (a) HH polarized monostatic RCS of the target in Fig. 3.6, (b) VV polarized monostatic RCS of the plate in Fig. 3.6. The BI portion of the FE–BI systems was formed using CFIE with the combination parameter α set to 0.5 as indicated

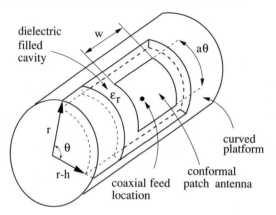

FIGURE 3.8: Microstrip patch antenna (3.5 cm × 3.5 cm) on a cylindrical surface of radius 14.95 cm. The patch is on 0.3175 cm thick substrate having $\varepsilon_r = 2.32$

3.5.2   Antenna Radiation

The flexibility of the finite element modeling allows the FE–BI method to be easily extended to radiation problems. As an example, consider a 3.5 cm × 3.5 cm cavity backed patch mounted on a cylindrical surface of radius 14.95 cm. The patch is placed on the aperture of a 0.3175 cm deep cavity filled with a dielectric layer having $\varepsilon_r = 2.32$. Fig. 3.8 shows the layout of the curved antenna obtained by wrapping the flat antenna on the cylindrical surface, which corresponds to that considered in [98].

As outlined in [98], two different radiation modes (axial and circumferential) exist in the operation of the patch depicted in Fig. 3.8, depending on the feed location. The radiation patterns for the axial and circumferential excitation modes are shown in Fig. 3.9(a). It is seen

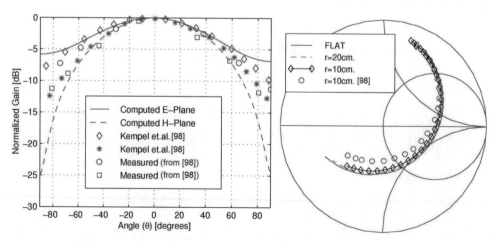

FIGURE 3.9: (a) Radiation pattern of the curved antenna in Fig. 3.8; (b) input impedance for the circumferentially polarized antenna in Fig. 3.8 (after Sertel and Volakis, © IEEE, 1998 [100])

that the FE–BI calculations are in agreement with measurements in the broadside region, but the agreement deteriorates toward the shadow directions (around $\theta = -90°$ and $\theta = 90°$). This is because the half-space Green's function [99, 100] was used for generating the data in Fig. 3.9, whereas in [98] the exact cylindrical Green's function was employed. In the region $-60° \leq \theta \leq 60°$, the measured and calculated patterns agree to within 1 dB. The input impedance dependence on antenna curvature is also plotted in Fig. 3.9(b) for the circumferentially polarized patch. These values are in good agreement to those presented in [98]. Further, the resonant frequency can be accurately predicted since it is a more local phenomenon.

In Fig. 3.10 we show example calculations for a 5-element patch array on a curved surface. The geometrical configuration of the array is shown in Fig. 3.10(a). Since a closed form Green's function for arbitrary doubly curved structures does not exist, one may use an approximate Green's function. Here, the half-space Green's function is used as if the cavity resides in a flat ground plane. It is demonstrated that this approach is applicable for geometries which are not highly curved.

We define the array geometry starting from a flat array with square patch antennas of size 3.5 cm × 2.625 cm. Two separate cases are considered for two different element spacings. For the first case, the flat geometry is backed by a rectangular cavity of dimensions 33.25 cm × 6.125 cm × 0.3175 cm; for the second case, the cavity dimensions are 28.0 cm × 6.125 cm × 0.3175 cm. The separation between the elements was 6.125 cm and 5.25 cm for the two cases, respectively. In both cases, the array is recessed in an infinite metallic ground plane. The cavities are filled with a material of relative permittivity 2.32. Curved arrays are then formed from the flat arrays by wrapping them onto metallic circular cylinders of radii 20 cm and 10 cm,

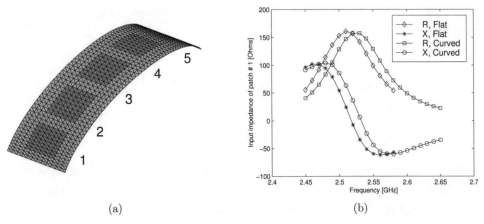

(a)   (b)

FIGURE 3.10: (a) Curved patch array geometry; (b) input impedance for patch element 1 (radius of curvature = 20 cm)

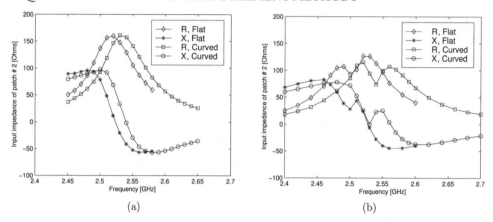

FIGURE 3.11: (a) Input impedance for patch element 2 (radius of curvature = 20 cm); (b) input impedance for patch element 2 (radius of curvature = 10 cm)

respectively. All array elements were fed in-phase by an offset vertical probe of constant current. The surface mesh used in the FE–BI solutions is given as in Fig. 3.10(a) (for the curved array with spacing 6.125 cm).

A comparison of the input impedances of the individual patches on the flat and curved surface (with a 10 cm radius of curvature) is shown in Figs. 3.10(b) and 3.11. The increased coupling between the different patches due to curvature can be especially seen in the input impedance of patch 2.

The FE–BI approach was recently extended to model very large antenna arrays using the translational symmetries of the array geometry. When dealing with arrays, certain advantages associated with the repeatability of each array element can be exploited [93, 101]. More specifically, we observe from Fig. 3.12 that each element has identical geometry and thus the FE–BI matrix system takes the form

$$
\begin{bmatrix}
\mathbf{a}_{11} & \mathbf{a}_{12} & \cdots & \mathbf{a}_{1M} \\
\mathbf{a}_{21} & \mathbf{a}_{22} & \cdots & \mathbf{a}_{2M} \\
\vdots & \vdots & \ddots & \vdots \\
\mathbf{a}_{M1} & \mathbf{a}_{M2} & \cdots & \mathbf{a}_{MM}
\end{bmatrix}
\begin{Bmatrix}
\mathbf{x}_1 \\
\mathbf{x}_2 \\
\vdots \\
\mathbf{x}_M
\end{Bmatrix}
=
\begin{Bmatrix}
\mathbf{b}_1 \\
\mathbf{b}_2 \\
\vdots \\
\mathbf{b}_M
\end{Bmatrix},
\tag{3.34}
$$

where the submatrices $\mathbf{a}_{mm'}$ denote the individual coupling operators between the m and m' elements in the array, $\{\mathbf{x}_1, \ldots, \mathbf{x}_M\}^T$ is a block-vector containing the unknown vectors for each array element, and $\{\mathbf{b}_1, \ldots, \mathbf{b}_M\}^T$ is also a block-vector providing the excitations on the array elements. Each of the diagonal submatrices is of the same form as given in Eq. (3.22), whereas the off-diagonal submatrices describe the coupling among the m and m' array elements (see

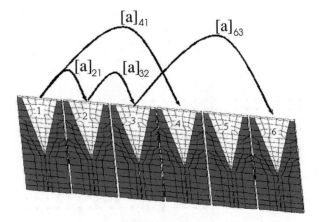

FIGURE 3.12: Illustration of redundant coupling paths in a 1×6 array

Fig. 3.12). If we use a boundary integral to enclose the volume of each array element, then all off-diagonal submatrices will just contain the $\bar{Z}^{JM}$ and $\bar{Z}^{JJ}$ submatrices in Eq. (3.22).

Of importance is that the coupling submatrices only depend on the absolute distance among elements and the diagonal submatrices are identical. Consequently, the overall matrix in Eq. (3.34) is block Toeplitz. Thus, Eq. (3.34) can be cast in the circulant form $\Pi * \{x\} = \{b\}$ where $*$ implies convolution and $\Pi = \{a_{M-1} \ldots a_1/, a_0/, a_{-1} \ldots a_{1-M}\}$ in which notation $a_{mm'} = a_{m-m'} = a_p$ has been introduced. This observation implies that the fast Fourier transform (FFT) can be employed to carry out the matrix-vector products of the entire FE–BI system in $O(N \log N)$ CPU time and using $O(N)$ storage.

In addition to the above, we may further exploit the fact that all diagonal submatrices, i.e., $a_{mm'} = a_{m-m'} = a_p$ for $p = 0$, are identical, and use the inverse of a_0 to precondition the entire matrix system. This type of preconditioning has been found very effective once the overall FE–BI matrix is restructured as suggested in Eq. (3.34). Details of this decomposition method can be found in [93].

We conclude this section by presenting the radiation of a large $30 \times 30 = 900$ element rectangular array of tapered slot antennas (TSAs) (see Fig. 3.13 for the element geometry). The discretization of this element resulted in 1103 unknowns (506 FE and 596 BI unknowns) leading to nearly a million unknowns for the entire matrix system (992,700 unknowns). By using the proposed decomposition method illustrated in Fig. 3.12, the storage alone was reduced from 3.8 TBytes down to 16 GBytes, and could thus be solved. Moreover, the solution time was only 18.6 hours, a reduction by three orders of magnitude as compared to a conventional FE–BI implementation. Thus, realistic finite arrays can be evaluated and designed using the proposed decomposition method. As an example, Fig. 3.14 shows the field distribution and corresponding array pattern for a 16×16 TSA array with 20 dB Taylor tapering ($\bar{n} = 4$).

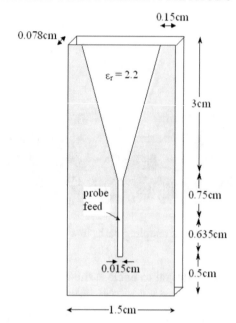

FIGURE 3.13: Tapered slot antenna element

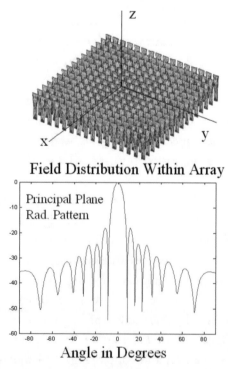

FIGURE 3.14: Field distribution and radiation pattern of a 16×16 TSA array with 20 dB Taylor tapering ($\bar{n} = 4$)

The hybrid FE–BI method is a very powerful electromagnetic field calculation tool. Until recently, the large CPU time and memory complexity of the BI restricted the FE–BI method to moderately sized discretization models. With the introduction of fast methods such as the fast multipole method, the applicability of the FE–BI can greatly be extended beyond the limitations of direct solution methods. Among the fast methods, adaptive integral method (AIM) has also proven very attractive if planar BI termination surfaces are involved. In this case, only 2D FFTs need be performed and periodic Green's function can be included in a straightforward manner since they have the required convolutional property. On the other hand, the concept of FMM is basically three dimensional and, therefore, FMM is favorably used for nonplanar termination surfaces. Both AIM and FMM attain their speedups through an efficient computation of the BI far interactions.

A further improvement of the FE–BI for modeling the periodic structure using an efficient evaluation the of the periodic Green's function series by virtue of the Ewald transformation is also discussed in [102].

CHAPTER 4

Hybrid Volume–Surface Integral Equation

In the previous two chapters, we saw an effective hybrid method for modeling arbitrarily shaped inhomogeneous composite structures, namely the FE–BI method. When a fast integral equation solver—such as the multilevel fast multipole method (MLFMM) [103], the adaptive integral method (AIM) [104], or singular value decomposition based methods (SVD) [22]—is employed to combat the $O(N^2)$ memory and computational complexity of the MoM solution of the BI, the FE–BI method becomes one of the most versatile and efficient approaches in computational electromagnetics. However, the FE–BI technique does have some disadvantages. Specifically, for a given $\mathbf{E}$ field or $\tilde{\mathbf{H}}$ field FE–BI formulation, satisfaction of all the boundary conditions discussed in Section 2.1.3, especially the sheet boundary condition, leaves much to be desired. Furthermore, for high contrast materials (i.e., materials with large permittivity and/or permeability values), this hybrid approach can become inefficient and can produce badly conditioned system matrices. Thus, any savings obtained due to the $O(N)$ complexity of the FE approach is wasted due to nonconvergent iterative solutions. In addition, for cases where one is simulating very thin homogeneous materials, the FE–BI system essentially becomes dominated by the BI equations. On the other hand, for any simulation of large volumes of homogeneous material, the FE solution requires that the whole volume be discretized causing the number of unknowns to be unnecessarily large.

One promising approach to combat the drawbacks of the hybrid FE–BI method, while still maintaining the arbitrariness of the structures being modeled, was presented by the authors in [105] called the *generalized volume–surface integral equation* (VSIE) method. Though we note that a similar method, unbeknown to the authors at the time, was first proposed by Bleszynski in [106]. Interestingly, both works were accomplished trying to solve the same problem for two distinct applications, namely modeling high contrast material objects. The work in [106] was motivated by efficient modeling of the highly complex biological tissues found in medical imaging applications, while the work in [105] was motivated by the modeling of highly complex metamaterial structures [107].

Surface integral equation (SIE) methods based on the PMCHWT (Poggio, Miller, Chang, Harrington, Wu, Tsai) formulation [108–110] are quite resilient (especially when far-field information is required) in modeling high contrast materials since homogeneous material domains can be represented by appropriate Green's functions. In addition, it is possible to generalize the SIE to handle arbitrary piecewise homogeneous materials through the use of an appropriate current junction resolution process [111–117]. Because domain discretization only occurs on the boundaries separating piecewise homogeneous materials and because practically any boundary condition can be satisfied across these boundaries, we can already expect the generalized SIE formulation itself to outperform FE–BI formulations for the specific cases discussed above. However, the FE–BI still allows one the flexibility to model inhomogeneous and/or anisotropic materials. This deficiency is solved by incorporating the volume integral equation (VIE) [1, 33, 39, 118–122] into the generalized SIE formulation [105]. This is accomplished through a factorization of the material parameters that alleviates the computational burden associated with typical VIE implementations. As such, the VIE is only invoked over regions where the material is actually varying, and the SIE is employed elsewhere along the boundaries of the homogeneous (possibly high contrast) regions. Further, since the VIE is only used to account for material perturbations, there is a significant reduction in the discretization rate often associated with the simulation of high contrast materials [123].

In the following, we begin by developing the generalized VSIE starting with a unique decomposition of the material parameters, ε and μ. This is followed by the MoM solution using the curvilinear hexahedral and quadrilateral finite elements discussed in Section 1.3. Also, a proper set of testing functions [24] is presented for the VIE which results from the Petrov–Galerkin variational statement from Section 1.1.2. Through the use of a junction resolution algorithm [114, 116, 117], we then discuss the matrix assembly process and demonstrate the efficiency and accuracy of the VSIE formulation.

4.1   GENERALIZED VSIE FORMULATION

In this section, we formulate the generalized VSIE based on the application of Green's method within individual regions of the volumetric partitioning shown in Fig. 4.1. Although the use of Green's identity to derive integral equation formulations is pretty standard electromagnetic theory [122, 124–126], we explicitly state the steps of Green's method in order to elucidate an interesting result pertaining to the simulation of composite materials through the use of a single field unknown that mirrors the formulation arrived in [126, 127].

To describe the generalized VSIE, we refer back to Fig. 4.1 which shows several volume sections (V_0, V_1, . . . , $V_{N_{vol}}$), each associated with a material medium which may be associated with either a uniform or a varying material profile. Because of the latter, it is important to

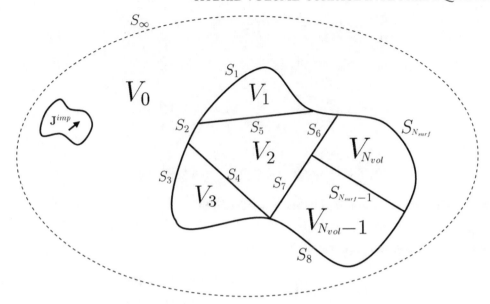

FIGURE 4.1: Partitioning of space for the generalized VSIE

represent the relative permittivity in individual volumes as

$$\bar{\bar{\varepsilon}}_r(\mathbf{r}) = \varepsilon_b \bar{\bar{I}} + \bar{\bar{\varepsilon}}_\delta(\mathbf{r}) = \varepsilon_b [\bar{\bar{I}} + \varepsilon_b^{-1} \bar{\bar{\varepsilon}}_\delta(\mathbf{r})] = \varepsilon_b \bar{\bar{\varepsilon}}_\Delta(\mathbf{r}), \tag{4.1}$$

in which $\bar{\bar{I}}$ is the identity dyad. Similarly, the permeability $\bar{\bar{\mu}}_r$ can be written as

$$\bar{\bar{\mu}}_r(\mathbf{r}) = \mu_b \bar{\bar{I}} + \bar{\bar{\mu}}_\delta(\mathbf{r}) = \mu_b [\bar{\bar{I}} + \mu_b^{-1} \bar{\bar{\mu}}_\delta(\mathbf{r})] = \mu_b \bar{\bar{\mu}}_\Delta(\mathbf{r}). \tag{4.2}$$

Here, the uniform terms ε_b and μ_b constitute an equivalent background homogeneous region that can be varied by way of modulation terms $\bar{\bar{\varepsilon}}_\Delta(\mathbf{r})$ and $\bar{\bar{\mu}}_\Delta(\mathbf{r})$ to obtain (4.1) and (4.2). In what follows, we will see that it is these modulation terms that contribute to the "equivalent volume sources" within the generalized VSIE.

As shown in Fig. 4.1, we can partition the entire volume V into $(N_{vol} + 1)$ volumes so that $V = \bigcup_{i=0}^{N_{vol}} V_i$ and $V_i \cap V_j = \{\emptyset\}$ for $i \neq j$. Each volume V_i is characterized by the parameter set $\{\varepsilon_b^i, \bar{\bar{\varepsilon}}_\Delta^i, \mu_b^i, \bar{\bar{\mu}}_\Delta^i\}$ with wavenumber k_i given by $k_i = k_0 \sqrt{\varepsilon_b^i \mu_b^i}$. Volume V_0 is assumed here to be free space with wavenumber k_0. We denote the boundary surfaces defined by the adjoining volumes as S_j, $\{j = 1 \ldots N_{surf}\}$, where the number of surfaces, N_{surf}, depends on the topology of the partitioned volumes. The complete boundary surrounding V_i can be represented as $\partial V_i = \bigcup_{j \in \{p_i\}} S_j$, where set $\{p_i\}$ contains the numbers of surfaces enclosing V_i. Since each surface S_j is a boundary formed by the junction of the volume pair $(V_{S_j}^+, V_{S_j}^-)$, where $V_{S_j}^\pm \in \{V_i\}$,

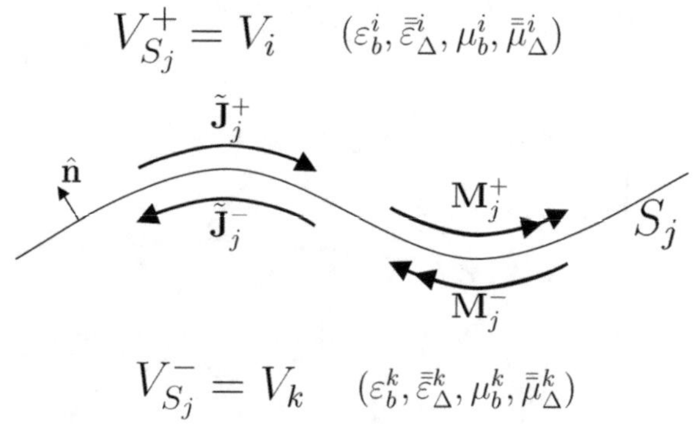

$$V_{S_j}^+ = V_i \quad (\varepsilon_b^i, \bar{\bar{\varepsilon}}_\Delta^i, \mu_b^i, \bar{\bar{\mu}}_\Delta^i)$$

$$V_{S_j}^- = V_k \quad (\varepsilon_b^k, \bar{\bar{\varepsilon}}_\Delta^k, \mu_b^k, \bar{\bar{\mu}}_\Delta^k)$$

FIGURE 4.2: Current relationships for the generalized VSIE

we introduce the convenient symbolic function

$$\text{sign}(i, j) = \begin{cases} + & \text{if} \quad V_i = V_{S_j}^+ \\ - & \text{if} \quad V_i = V_{S_j}^-. \end{cases} \tag{4.3}$$

Throughout this chapter, we will also assume that all normals $\hat{n}$ on any surface S_j point in the direction of volume $V_{S_j}^+$ (see Fig. 4.2).

With the above definitions and notation, let us begin the development of the VSIE by turning our attention to the ith volume partition V_i that is enclosed by boundary ∂V_i. Everywhere within this volume, Maxwell's equations must be satisfied and are stated here for time-harmonic electric and magnetic fields

$$\nabla \times \mathbf{E}_i = -jk_0\mu_b^i\bar{\bar{\mu}}_\Delta^i \cdot \tilde{\mathbf{H}}_i, \tag{4.4}$$

$$\nabla \times \tilde{\mathbf{H}}_i = jk_0\varepsilon_b^i\bar{\bar{\varepsilon}}_\Delta^i \cdot \mathbf{E}_i + \tilde{\mathbf{J}}_i^{\text{imp}}, \tag{4.5}$$

where an $e^{j\omega t}$ time dependence is assumed. The augmented magnetic field $\tilde{\mathbf{H}}_i$ is defined with respect to the free-space impedance Z_0 to be $\tilde{\mathbf{H}}_i \triangleq Z_0\mathbf{H}_i$ and the augmented source term $\tilde{\mathbf{J}}_i^{\text{imp}} = Z_0\mathbf{J}_i^{\text{imp}}$ represents any impressed current sources. It is straightforward to show that the system of equations in (4.4) and (4.5) result in the following vector wave equation for the electric field intensity:

$$\nabla \times \nabla \times \mathbf{E}_i - k_i^2\mathbf{E} = k_i^2(\bar{\bar{\varepsilon}}_\Delta^i - \bar{\bar{I}}) \cdot \mathbf{E}_i - jk_0\mu_b^i\nabla \times [(\bar{\bar{\mu}}_\Delta^i - \bar{\bar{I}}) \cdot \tilde{\mathbf{H}}_i] - jk_0\mu_b^i\tilde{\mathbf{J}}_i^{\text{imp}}. \tag{4.6}$$

As is typical in any integral equation derivation using Green's method, to solve the vector wave equation in (4.6), we consider Green's dyadic

$$\bar{\bar{G}}_{k_i}(\mathbf{r}, \mathbf{r}') = [\bar{\bar{I}} + \frac{1}{k_i^2}\nabla\nabla]g_{k_i}(\mathbf{r}, \mathbf{r}'), \tag{4.7}$$

which satisfies

$$\nabla \times \nabla \times \bar{\bar{G}}_{k_i}(\mathbf{r}, \mathbf{r}') - k_i^2 \bar{\bar{G}}_{k_i}(\mathbf{r}, \mathbf{r}') = \bar{\bar{I}}\delta(\mathbf{r} - \mathbf{r}'). \tag{4.8}$$

Here the scalar Green's function $g_{k_i}(\mathbf{r}, \mathbf{r}')$ is given by

$$g_{k_i}(\mathbf{r}, \mathbf{r}') = \frac{e^{-jk_i|\mathbf{r}-\mathbf{r}'|}}{4\pi|\mathbf{r} - \mathbf{r}'|}. \tag{4.9}$$

If we now apply the inner product of (4.6) with $\bar{\bar{G}}_{k_i}$ and (4.8) with $\mathbf{E}_i$, subtract the resulting two equations, and integrate over the whole volume V_i, then we can arrive at the following expression,

$$\mathbf{E}_i(\mathbf{r}) + \int_{V_i} [\nabla' \times \nabla' \times \mathbf{E}_i \cdot \bar{\bar{G}}_{k_i} - \mathbf{E}_i \cdot \nabla' \times \nabla' \times \bar{\bar{G}}_{k_i}] dr'$$
$$= k_i^2 \int_{V_i} \bar{\bar{G}}_{k_i} \cdot (\bar{\bar{\varepsilon}}_\Delta^i - \bar{\bar{I}}) \cdot \mathbf{E}_i \, dr' - jk_0\mu_b^i \int_{V_i} \bar{\bar{G}}_{k_i} \cdot \nabla' \times [(\bar{\bar{\mu}}_\Delta^i - \bar{\bar{I}}) \cdot \tilde{\mathbf{H}}_i] dr' + \mathbf{E}^{\text{inc}}(\mathbf{r})\delta[i], \tag{4.10}$$

where for scattering applications $\mathbf{E}^{\text{inc}}$ is the electric field due to the impressed current source $\tilde{\mathbf{J}}_0^{\text{imp}}$ placed somewhere within the free-space region V_0 (see Fig. 4.1), and $\delta[i]$ is the standard Kronecker delta-function. To simplify the expressions in (4.10), it is typical to make use of Green's identity [124], namely,

$$\int_V [\nabla \times \nabla \times \mathbf{P}(\mathbf{r}) \cdot \bar{\bar{Q}}(\mathbf{r}, \mathbf{r}') - \mathbf{P}(\mathbf{r}) \cdot \nabla \times \nabla \times \bar{\bar{Q}}(\mathbf{r}, \mathbf{r}')] dr$$
$$= \oint_{\partial V} [(\hat{\mathbf{n}} \times \mathbf{P}(\mathbf{r})) \cdot \nabla \times \bar{\bar{Q}}(\mathbf{r}, \mathbf{r}') + (\hat{\mathbf{n}} \times \nabla \times \mathbf{P}(\mathbf{r})) \cdot \bar{\bar{Q}}(\mathbf{r}, \mathbf{r}')] dr, \tag{4.11}$$

where $\mathbf{P}$ is any well-behaved vector field, $\bar{\bar{Q}}$ is any well-behaved dyadic operator, ∂V is the boundary surface containing V, and $\hat{\mathbf{n}}$ is the outward normal to surface ∂V. Applying (4.11) to (4.10) and substituting (4.4) into the resulting expression yields

$$\mathbf{E}_i(\mathbf{r}) + \sum_{j=\{p_i\}} \text{sign}(i, j) \int_{S_j} [\nabla' g_{k_i} \times (\hat{\mathbf{n}}' \times \mathbf{E}_i) + jk_0\mu_b^i \bar{\bar{G}}_{k_i} \cdot (\hat{\mathbf{n}}' \times (\bar{\bar{\mu}}_\Delta^i \cdot \tilde{\mathbf{H}}_i))] dr'$$
$$= k_i^2 \int_{V_i} \bar{\bar{G}}_{k_i} \cdot (\bar{\bar{\varepsilon}}_\Delta^i - \bar{\bar{I}}) \cdot \mathbf{E}_i \, dr' - jk_0\mu_b^i \int_{V_i} \bar{\bar{G}}_{k_i} \cdot \nabla' \times [(\bar{\bar{\mu}}_\Delta^i - \bar{\bar{I}}) \cdot \tilde{\mathbf{H}}_i] dr' + \mathbf{E}^{\text{inc}}(\mathbf{r})\delta[i], \tag{4.12}$$

where identity $\mathbf{P} \cdot \nabla' \times \bar{\bar{G}}_{k_i} = -\nabla' g_{k_i} \times \mathbf{P}$ has been utilized. Let us now turn our attention to the second integrand of the surface integral in (4.12). Because we assumed the presence of arbitrary composite materials, a rather awkward expression resulted, namely, $\hat{\mathbf{n}}' \times (\bar{\bar{\mu}}_\Delta^i \cdot \tilde{\mathbf{H}}_i)$. At first glance, when anisotropic magnetic materials are considered, it does not appear possible to define the usual surface equivalent sources on the boundary ∂V_i as is done in typical PMCHWT formulations [114], which assume isotropic composite materials. However, if we now turn our attention to the second volume integral in (4.12), it is possible to prove using integration by parts that

$$
\int_{V_i} \bar{\bar{G}}_{k_i} \cdot \nabla' \times [(\bar{\bar{\mu}}_\Delta^i - \bar{\bar{I}}) \cdot \tilde{\mathbf{H}}_i] dr' = -\int_{V_i} \nabla' g_{k_i} \times [(\bar{\bar{\mu}}_\Delta^i - \bar{\bar{I}}) \cdot \tilde{\mathbf{H}}_i] dr'
$$
$$
- \sum_{j=\{p_i\}} \text{sign}(i,j) \int_{S_j} \bar{\bar{G}}_{k_i} \cdot [\hat{\mathbf{n}}' \times ((\bar{\bar{\mu}}_\Delta^i - \bar{\bar{I}}) \cdot \tilde{\mathbf{H}}_i)] dr', \tag{4.13}
$$

and after substitution of (4.13) into (4.12), we have the resulting integral equation

$$
\mathbf{E}_i(\mathbf{r}) + \sum_{j=\{p_i\}} \text{sign}(i,j) \int_{S_j} [\nabla' g_{k_i} \times (\hat{\mathbf{n}}' \times \mathbf{E}_i) + jk_0 \mu_b^i \bar{\bar{G}}_{k_i} \cdot (\hat{\mathbf{n}}' \times \tilde{\mathbf{H}}_i)] dr'
$$
$$
- k_i^2 \int_{V_i} \bar{\bar{G}}_{k_i} \cdot (\bar{\bar{\varepsilon}}_\Delta^i - \bar{\bar{I}}) \cdot \mathbf{E}_i \, dr' - jk_0 \mu_b^i \int_{V_i} \nabla' g_{k_i} \times [(\bar{\bar{\mu}}_\Delta^i - \bar{\bar{I}}) \cdot \tilde{\mathbf{H}}_i] dr' = \mathbf{E}^{\text{inc}}(\mathbf{r}) \delta[i], \tag{4.14}
$$

where the kernels containing the equivalent volume sources have been moved to the left-hand side. If on each surface $\{S_j\}$ (see Fig. 4.2) we define a set of four equivalent currents denoted by $\tilde{\mathbf{J}}_j^+, \tilde{\mathbf{J}}_j^-, \mathbf{M}_j^+$, and $\mathbf{M}_j^-$ such that

$$
\tilde{\mathbf{J}}_j^\pm = \pm \hat{\mathbf{n}} \times \tilde{\mathbf{H}}_j^\pm, \quad \text{and} \quad \mathbf{M}_j^\pm = \mp \hat{\mathbf{n}} \times \mathbf{E}_j^\pm, \tag{4.15}
$$

and since

$$
\tilde{\mathbf{J}}_j^{\text{sign}(i,j)} = \text{sign}(i,j)(\hat{\mathbf{n}} \times \tilde{\mathbf{H}}_i), \tag{4.16}
$$
$$
\mathbf{M}_j^{\text{sign}(i,j)} = -\text{sign}(i,j)(\hat{\mathbf{n}} \times \mathbf{E}_i), \tag{4.17}
$$

we can define $\tilde{\mathbf{J}}_i \triangleq \sum_{j=\{p_i\}} \tilde{\mathbf{J}}_j^{\text{sign}(i,j)}$ and $\mathbf{M}_i \triangleq \sum_{j=\{p_i\}} \mathbf{M}_j^{\text{sign}(i,j)}$ and rewrite (4.14) as

$$
\mathbf{E}_i(\mathbf{r}) - \mathbf{E}_i^{\text{scat}}(\mathbf{r}) = \mathbf{E}^{\text{inc}}(\mathbf{r}) \delta[i], \tag{4.18}
$$

where

$$
\mathbf{E}_i^{\text{scat}} = \Psi_i(\bar{\bar{\varepsilon}}_\Delta^i, \mathbf{E}_i) + \Omega_i(\bar{\bar{\mu}}_\Delta^i, \nabla \times \mathbf{E}_i) + \Phi_j^i(\mathbf{M}_i) - \mu_b^i \Theta_j^i(\tilde{\mathbf{J}}_i), \tag{4.19}
$$

and

$$
\Phi_j^i(\mathbf{Y}) = \int_{S_j} \nabla' g_{k_i}(\mathbf{r}, \mathbf{r}') \times \mathbf{Y}(\mathbf{r}') \, dr', \tag{4.20}
$$

$$
\Theta_j^i(\mathbf{Y}) = jk_0 \int_{S_j} g_{k_i}(\mathbf{r}, \mathbf{r}') \mathbf{Y}(\mathbf{r}') \, dr'
$$
$$
- \frac{j}{k_0} \nabla \int_{S_j} g_{k_i}(\mathbf{r}, \mathbf{r}') \nabla' \cdot \mathbf{Y}(\mathbf{r}') \, dr', \tag{4.21}
$$

$$\Omega_i(\bar{\bar{\alpha}}, \mathbf{X}) = \int_{V_i} \nabla' g_{k_i}(\mathbf{r}, \mathbf{r}') \times [\bar{\bar{\alpha}}^{-1}(\mathbf{r}') - \bar{\bar{I}}] \cdot \mathbf{X}(\mathbf{r}') \, dr', \qquad (4.22)$$

$$\Psi_i(\bar{\bar{\alpha}}, \mathbf{X}) = k_i^2 \int_{V_i} g_{k_i}(\mathbf{r}, \mathbf{r}')[\bar{\bar{\alpha}}(\mathbf{r}') - \bar{\bar{I}}] \cdot \mathbf{X}(\mathbf{r}') \, dr'$$
$$- \nabla \int_{V_i} \nabla' g_{k_i}(\mathbf{r}, \mathbf{r}') \cdot [\bar{\bar{\alpha}}(\mathbf{r}') - \bar{\bar{I}}] \cdot \mathbf{X}(\mathbf{r}') \, dr'. \qquad (4.23)$$

Note that integral operators Φ_j^i and Θ_j^i represent mappings from $\mathcal{H}^{-1/2}(\mathrm{Div}; \partial V_i)$ to $\mathcal{H}(\mathrm{Curl}; V_i)$ and operators Ω_i and Ψ_i represent mappings from $\mathcal{H}(\mathrm{Curl}; V_i)$ to $\mathcal{H}(\mathrm{Curl}; V_i)$. Since (4.18) has three electromagnetic quantities as unknowns, it is necessary to provide two more equations so that a determined system is obtained for MoM implementations. The first of these two equations essentially is the restriction of the observation point in (4.18) on boundary ∂V_i through the tangential trace operator γ_t. This produces the following surface integral equation,

$$\frac{1}{2}\gamma_t \mathbf{E}_i(\mathbf{r}) - \gamma_t \mathbf{E}_i^{\mathrm{scat}}(\mathbf{r}) = \gamma_t \mathbf{E}_i^{\mathrm{inc}}(\mathbf{r})\delta[i], \qquad (4.24)$$

where we note that $\gamma_t \mathbf{E}_i = \sum_{j=\{p_i\}} \mathrm{sign}(i, j)\hat{n} \times \mathbf{M}_j^{\mathrm{sign}(i,j)}$ and that the integral in (4.20) has now become a principal value integral excluding the observation point $\mathbf{r}$. The final integral equation is essentially the dual of (4.24), where (4.4) and (4.5) are used to ensure that $\mathbf{E}_i$ (not $\tilde{\mathbf{H}}_i$) is the field unknown. This results in the following surface integral equation,

$$\frac{1}{2}\gamma_t \tilde{\mathbf{H}}_i(\mathbf{r}) - \gamma_t \tilde{\mathbf{H}}_i^{\mathrm{scat}}(\mathbf{r}) = \gamma_t \tilde{\mathbf{H}}_i^{\mathrm{inc}}(\mathbf{r})\delta[i], \qquad (4.25)$$

where $\gamma_t \tilde{\mathbf{H}}_i = -\sum_{j=\{p_i\}} \mathrm{sign}(i, j)\hat{n} \times \tilde{\mathbf{J}}_j^{\mathrm{sign}(i,j)}$ and $\tilde{\mathbf{H}}_i^{\mathrm{scat}}$ is given by

$$\tilde{\mathbf{H}}_i^{\mathrm{scat}}(\mathbf{r}) = -\mathrm{j}k_0 \varepsilon_b^i [\Psi_i(\bar{\bar{\mu}}_\Delta^{i}{}^{-1}, \nabla \times \mathbf{E}_i) + \Omega_i(\bar{\bar{\varepsilon}}_\Delta^{i}{}^{-1}, \mathbf{E}_i)] - \Phi_j^i(\tilde{\mathbf{J}}_i) - \varepsilon_b^i \Theta_j^i(\mathbf{M}_i). \qquad (4.26)$$

The complete generalized VSIE formulation consists of the integral equations expressed in (4.18), (4.24), and (4.25). This system actually represents three equations per partitioned volume. It is important to note, however, that these integral equations are coupled [116] through the satisfaction of the boundary conditions maintained between the partitioned volumes. This coupling is accomplished through application of a *junction resolution* algorithm [114, 116, 117], which is briefly described in the next section. Also, the electric field is the only unknown in those regions consisting of equivalent volume currents, i.e., for volumes with material properties differing from the background material of the partitioned volumes (such as anisotropic media). This flexibility can be very appealing for certain electromagnetic simulations.

Interestingly, when only one volume partition V_0 is considered, the VIE in (4.18) reduces to that discussed in [127] for the modeling of composite materials with only one field unknown.

This results from the steps taken to go from (4.12) to (4.14), where it was commonly believed that the surface integral kernel in (4.13) was also necessary to model permeable materials.

In the next section, we discuss the equivalent current relationships for different boundary conditions. Then we will present the variational form of the VSIE in (4.18), (4.24), and (4.25), and we will see that care must be taken when selecting the testing basis for the VIE in (4.18).

4.2   BOUNDARY CONDITIONS

We now take a closer look at the equivalent surface currents defined on the boundary surface S_j that separates volumes $V_{S_j}^+$ and $V_{S_j}^-$ (see Fig. 4.2). Specifically, we are interested in the relationships between $\tilde{\mathbf{J}}_j^+$, $\tilde{\mathbf{J}}_j^-$, $\mathbf{M}_j^+$, and $\mathbf{M}_j^-$ found by considering the material properties (therefore the boundary conditions defining S_j). We note that this approach is outlined in [128].

To begin, let us assume that S_j separates two material regions (i.e., $V_{S_j}^+ \neq V_{S_j}^-$). If the natural material boundary condition is satisfied across S_j, then from (2.12) and (4.15) we see that $\tilde{\mathbf{J}}_j^- = -\tilde{\mathbf{J}}_j^+$ and $\mathbf{M}_j^- = -\mathbf{M}_j^+$. If a thin conducting (PEC) sheet is placed between the two material regions, then $\mathbf{M}_j^+ = \mathbf{M}_j^- = 0$ and the electric surface currents remain unpaired. Correspondingly, if a thin magnetically conducting (PMC) sheet is placed between the two material regions, then $\tilde{\mathbf{J}}_j^+ = \tilde{\mathbf{J}}_j^- = 0$ and the magnetic surface currents remain unpaired. If a thin composite material sheet separates the two material regions, then from (2.19) we have that

$$\hat{n} \times \begin{bmatrix} \tilde{\mathbf{J}}_j^+ \\ \tilde{\mathbf{J}}_j^- \\ \mathbf{M}_j^+ \\ \mathbf{M}_j^- \end{bmatrix} = \begin{bmatrix} -\frac{1}{4R_e} & 0 & 0 & 0 \\ 0 & \frac{1}{4R_e} & 0 & 0 \\ 0 & 0 & \frac{1}{4R_m} & 0 \\ 0 & 0 & 0 & -\frac{1}{4R_m} \end{bmatrix} \begin{bmatrix} 0 & 0 & \alpha & \beta \\ 0 & 0 & \beta & \alpha \\ \alpha & \beta & 0 & 0 \\ \beta & \alpha & 0 & 0 \end{bmatrix} \begin{bmatrix} \tilde{\mathbf{J}}_j^+ \\ \tilde{\mathbf{J}}_j^- \\ \mathbf{M}_j^+ \\ \mathbf{M}_j^- \end{bmatrix}, \quad (4.27)$$

where $\alpha = 4R_e R_m + 1$ and $\beta = 4R_e R_m - 1$. Also note that it is possible to incorporate an additional coupling term to this sheet boundary condition that can effectively model scattering from thin material layers [74]. The above boundary condition is easily incorporated in the SIEs in (4.24) and (4.25) where (4.27) can be substituted directly into the terms containing $\hat{n} \times \tilde{\mathbf{J}}_j^{\text{sign}(i,j)}$ and $\hat{n} \times \mathbf{M}_j^{\text{sign}(i,j)}$. For the special case when $R_m \to \infty$ or $\mu_r' = 1$, we then set $\mathbf{M}_j^- = -\mathbf{M}_j^+$. Also, when $R_e \to \infty$ or $\varepsilon_r' = 1$, we set $\tilde{\mathbf{J}}_j^- = -\tilde{\mathbf{J}}_j^+$.

Now let us consider the case when $V_{S_j}^-$ is a conducting medium. Obviously, if $V_{S_j}^-$ is a perfect electric conductor, then $\tilde{\mathbf{J}}_j^- = \mathbf{M}_j^+ = \mathbf{M}_j^- = 0$. Alternatively, if $V_{S_j}^-$ is a perfect magnetic conductor, then $\mathbf{M}_j^- = \tilde{\mathbf{J}}_j^+ = \tilde{\mathbf{J}}_j^- = 0$. For the case when $V_{S_j}^-$ is an imperfect conductor or a material-coated conductor, then from (2.15) we can set $\tilde{\mathbf{J}}_j^- = \mathbf{M}_j^- = 0$ and force $\mathbf{M}_j^+ = -\eta \hat{n} \times \tilde{\mathbf{J}}_j^+$.

Finally, we consider the case when $V_{S_j}^+ = V_{S_j}^-$. Most notably, for this case, we do not need to consider two-sided currents as long as the surface is not a thin material sheet. Thus, we can set $\tilde{\mathbf{J}}_j^- = \mathbf{M}_j^- = 0$ and assume that $\tilde{\mathbf{J}}_j^+ = \tilde{\mathbf{J}}_j$ and $\mathbf{M}_j^+ = \mathbf{M}_j$, where similar cancellations as before are undertaken for cases when the surface is a perfect or imperfect conductor. For the case when the surface is indeed a sheet boundary, one performs the same procedure as if $V_{S_j}^+ \neq V_{S_j}^-$.

4.3   VARIATIONAL FORM OF THE VSIE

In this section, we develop the variational form of the integral equation system described by (4.18), (4.24), and (4.25) that is valid within each partitioned volume V_i. To begin this development, we explicitly state the function spaces that contain the unknowns within the formulation. Similar to the FE–BI formulations discussed in Chapters 2 and 3, the electric field unknown $\mathbf{E}_i$ is contained in the vector space $\mathcal{W}_E$ defined as

$$\mathcal{W}_E^i = \{\mathbf{x} \in \mathcal{H}(\text{Curl}; V_i) \; : \; \gamma_t \mathbf{x}|_{\partial V_i \cap \Gamma_{\text{pec}}} = 0\}, \tag{4.28}$$

and the surface current unknowns $\tilde{\mathbf{J}}_i$ and $\mathbf{M}_i$ are contained in vector spaces $\mathcal{V}_J^i$ and $\mathcal{V}_M^i$, respectively, which are defined here as

$$\mathcal{V}_J^i = \{\mathbf{x} \in \mathcal{H}^{-1/2}(\text{Div}; \partial V_i) \; : \; \mathbf{x}|_{\partial V_i \cap \Gamma_{\text{pmc}}} = 0\}, \tag{4.29}$$

and

$$\mathcal{V}_M^i = \{\mathbf{x} \in \mathcal{H}^{-1/2}(\text{Div}; \partial V_i) \; : \; \mathbf{x}|_{\partial V_i \cap \Gamma_{\text{pec}}} = 0\}. \tag{4.30}$$

Before we express the variational statement of the VSIE, we first point out that the volume integral equation in (4.18) is actually an integral equation of the second kind because the field unknown is both inside and outside of the integral operator. Thus, Petrov–Galerkin's method discussed in Section 1.1.2 must be used for accurate variational treatment of (4.18). This requires that the testing space used in the variational treatment actually be the dual of $\mathcal{W}_E$; a fact that until recently [24] has been overlooked.

We are now ready for the variational statement. That is, we seek $\mathbf{E}_i \in \mathcal{W}_E$, $\mathbf{M}_i \in \mathcal{V}_M^i$, and $\tilde{\mathbf{J}}_i \in \mathcal{V}_J^i$ such that

$$\int_{V_i} \mathbf{t} \cdot \mathbf{E}_i \, dr - \int_{V_i} \mathbf{t} \cdot [\Psi_i(\bar{\bar{\varepsilon}}_\Delta^i, \mathbf{E}_i) - \Omega_i(\bar{\bar{\mu}}_\Delta^i, \nabla \times \mathbf{E}_i)] \, dr$$
$$- \int_{V_i} \mathbf{t} \cdot \Phi_j^i(\mathbf{M}_i) \, dr + \int_{V_i} \mathbf{t} \cdot \mu_b^i \Theta_j^i(\tilde{\mathbf{J}}_i) \, dr = \int_{V_i \cap V_0} \mathbf{t} \cdot \mathbf{E}^{\text{inc}} \, dr, \tag{4.31}$$

$$\varepsilon_b^i \int_{\partial V_i} \mathbf{v}_m \cdot \Theta_j^i(\mathbf{M}_i) \, dr - \int_{\partial V_i} \mathbf{v}_m \cdot \left[\frac{1}{2} \sum_{j=\{p_i\}} \text{sign}(i,j)\hat{n} \times \tilde{\mathbf{J}}_j^{\text{sign}(i,j)} - \Phi_j^i(\tilde{\mathbf{J}}_i) \right] dr$$
$$+ jk_0\varepsilon_b^i \int_{\partial V_i} \mathbf{v}_m \cdot [\Psi_i(\bar{\bar{\mu}}_\Delta^{i\,-1}, \nabla \times \mathbf{E}_i) + \Omega_i(\bar{\bar{\varepsilon}}_\Delta^{i\,-1}, \mathbf{E}_i)] \, dr = \int_{\partial V_i \cap V_0} \mathbf{v}_m \cdot \tilde{\mathbf{H}}^{\text{inc}}, \tag{4.32}$$

$$\mu_b^i \int_{\partial V_i} \mathbf{v}_j \cdot \Theta_j^i(\tilde{\mathbf{J}}_i) \, dr + \int_{\partial V_i} \mathbf{v}_j \cdot \left[\frac{1}{2} \sum_{j=\{p_i\}} \mathrm{sign}(i,j) \hat{n} \times \mathbf{M}_j^{\mathrm{sign}(i,j)} - \Phi_j^i(\mathbf{M}_i) \right] \, dr$$

$$- \int_{\partial V_i} \mathbf{v}_j \cdot [\Psi_i(\bar{\bar{\varepsilon}}_\Delta^i, \mathbf{E}_i) + \Omega_i(\bar{\bar{\mu}}_\Delta^i, \nabla \times \mathbf{E}_i)] \, dr = \int_{\partial V_i \cap V_0} \mathbf{v}_j \cdot \mathbf{E}^{\mathrm{inc}} \, dr, \tag{4.33}$$

for all $\mathbf{t} \in (\mathcal{W}_E^i)'$, $\mathbf{v}_m \in \mathcal{V}_M^i$, and $\mathbf{v}_j \in \mathcal{V}_J^i$.

4.4   DISCRETIZATION

In the previous section, we stated the weak form of the generalized VSIE. As usual, the next step is to construct the discretization framework upon which our numerical implementation will be based. As was done in Chapter 3 for the 3D FE–BI, we again use curl conforming elements described in (1.77) as our discretization elements for the electric field unknown. Likewise, divergence conforming elements described in (1.75) will be used for the equivalent surface currents. We should recall that the electric field is the only unknown in those regions where $\bar{\bar{\varepsilon}}_\Delta \neq \bar{\bar{I}}$ or $\bar{\bar{\mu}}_\Delta \neq \bar{\bar{I}}$. That is, for piecewise homogeneous objects, the formulation is no different than the generalized PMCHWT SIE formulations commonly used [114, 116, 117].

To begin with, let us consider the surface current expansions for a partitioned geometry containing N_{surf} surface elements given by

$$\tilde{\mathbf{J}}^\pm(\mathbf{r}) = \sum_{j=1}^{N_{\mathrm{surf}}} \sum_{n=1}^{N_{\mathrm{quad}}^j} \sum_{p=1}^{4} J_{np}^{\pm,j} \mathbf{v}_p(\mathbf{r}(u,v)), \tag{4.34}$$

$$\mathbf{M}^\pm(\mathbf{r}) = \sum_{j=1}^{N_{\mathrm{surf}}} \sum_{n=1}^{N_{\mathrm{quad}}^j} \sum_{p=1}^{4} M_{np}^{\pm,j} \mathbf{v}_p(\mathbf{r}(u,v)), \tag{4.35}$$

where N_{quad}^j is the number of quadrilateral elements comprising surface S_j, and $J_{np}^{\pm,j}$ and $M_{np}^{\pm,j}$ are the current values corresponding to the pth local edge of each element. The basis functions $\mathbf{v}_p$ are the divergence conforming roof top functions defined in (1.75) that allow for normal field continuity across element edges. To ensure that these expansions are contained in vector space $\mathcal{V}_J^i$ and $\mathcal{V}_M^i$, one must carefully pair the unknowns in a way that ensures satisfaction of the boundary conditions characterizing each surface element as well as the continuity conditions across the surface elements' edges. These so-called edges are most often referred to as *junctions*, and the procedure which ensures that the above conditions are satisfied is referred to as the *junction resolution algorithm*. A junction resolution algorithm valid for the basis functions employed here is discussed in Section 4.4.1.

In addition, the electric field expansion for geometries consisting of N_{vol} partitioned volumes is given by

$$\mathbf{E}(\mathbf{r}) = \sum_{i=1}^{N_{\mathrm{vol}}} \sum_{n=1}^{N_{\mathrm{hex}}^i} \sum_{p=1}^{12} E_{np}^i \mathbf{w}_p(\mathbf{r}(u, v, w)), \qquad (4.36)$$

where N_{hex}^i is the number of hexahedral elements comprising volume V_i and E_{np}^i denote the field values corresponding to the pth local edge belonging to the nth hexahedral element in V_i. The basis functions $\mathbf{w}_p$ are the curl conforming edge basis functions discussed in (1.77) that allow explicit enforcement of tangential field continuity across element faces. Equivalent to the FE–BI case, the expansion in (4.36) is contained in $\mathcal{W}_E^i$ after (i) all unknowns defined in material regions having property $\bar{\bar{\varepsilon}}_\Delta = \bar{\bar{I}}$ and $\bar{\bar{\mu}}_\Delta = \bar{\bar{I}}$ are eliminated, (ii) all unknowns corresponding to edges defined on PEC surfaces are also eliminated, and (iii) all remaining unknowns that are defined on edges containing multiple unknowns are grouped into a single set of global unknowns (see Section 2.4).

As mentioned in the previous section, a set of testing functions that are contained in the dual space of $\mathcal{W}_E^i$ must also be defined. In Section 1.1.1, we saw that for a given a finite-dimensional vector space, its dual space can be constructed by choosing testing functions that satisfy the property defined in (1.4). In light of this fact and in light of the Kronecker delta relationship maintained between covariant and contravariant vectors (i.e., $\mathbf{a}_i \cdot \mathbf{a}^j = \delta_{ij}$ for $i, j = u, v, w$), we proposed in [24] that a valid set of testing functions that span the space $(\mathcal{W}_E^i)' \subset \mathcal{H}(\mathrm{Div}; V_i)$ is given by

$$\mathbf{t}_1 = (1+v)(1+w)\frac{1}{J_V}\mathbf{a}_u \quad \mathbf{t}_5 = (1+u)(1+w)\frac{1}{J_V}\mathbf{a}_v \quad \mathbf{t}_9 = (1+u)(1+v)\frac{1}{J_V}\mathbf{a}_w,$$

$$\mathbf{t}_2 = (1+v)(1-w)\frac{1}{J_V}\mathbf{a}_u \quad \mathbf{t}_6 = (1+u)(1-w)\frac{1}{J_V}\mathbf{a}_v \quad \mathbf{t}_{10} = (1+u)(1-v)\frac{1}{J_V}\mathbf{a}_w,$$

$$\mathbf{t}_3 = (1-v)(1+w)\frac{1}{J_V}\mathbf{a}_u \quad \mathbf{t}_7 = (1-u)(1+w)\frac{1}{J_V}\mathbf{a}_v \quad \mathbf{t}_{11} = (1-u)(1+v)\frac{1}{J_V}\mathbf{a}_w,$$

$$\mathbf{t}_4 = (1-v)(1-w)\frac{1}{J_V}\mathbf{a}_u \quad \mathbf{t}_8 = (1-u)(1-w)\frac{1}{J_V}\mathbf{a}_v \quad \mathbf{t}_{12} = (1-u)(1-v)\frac{1}{J_V}\mathbf{a}_w,$$

$$(4.37)$$

where each $\mathbf{t}_p$ corresponds to a $\mathbf{w}_p$ in (4.36) for each volume element used in the electric field discretization. To see that these functions are a subset of $\mathcal{H}(\mathrm{Div}; V_i)$, consider the fact that the divergence defined in (1.72) of each testing function in (4.37) is identically zero. This rather interesting fact can be exploited when evaluating the matrix entries for the VIE part of the VSIE.

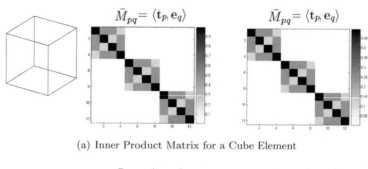

(a) Inner Product Matrix for a Cube Element

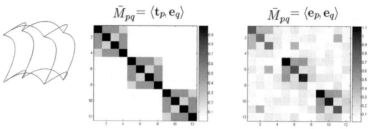

(b) Inner Product Matrix for a Distorted Cube Element

FIGURE 4.3: Illustration of the inner product matrix (4.38)

Another interesting property resulting from the application of the testing functions described in (4.37) is seen when we consider the 12×12 inner product matrix $\bar{M}$ defined by

$$\bar{M}_{pq} = \int_V \mathbf{h}_p \cdot \mathbf{w}_q \, dr, \qquad (4.38)$$

where $\mathbf{h}_p$ is either $\mathbf{t}_p$ or $\mathbf{w}_p$. We see in Fig. 4.3(a) that as a hexahedral element becomes distorted, the inner product matrix formed for the case when $\mathbf{h}_p = \mathbf{t}_p$ remains unperturbed. In contrast, for the case when $\mathbf{h}_p = \mathbf{w}_p$, as would be the case if typical Galerkin's testing were employed, we see in Fig. 4.3(b) that the resulting inner product matrix becomes less diagonally dominant. This often results in badly conditioned matrices since the inner product matrix is an actual kernel in the VIE variational statement (4.31). Hence, one can expect that solutions based upon typical Galerkin's testing, if not inaccurate, will at least be ill-conditioned for arbitrarily curved structures. Though we note that for rectilinear elements $\mathbf{t}_p = C \, \mathbf{w}_p$ for some constant C; which is probably the reason why this fact has been overlooked in the previous MoM implementations of the VIE.

4.4.1   Junction Resolution

At first glance, the current expansions in (4.34) and (4.35) appear to introduce a total of $16 N^j_{\text{quad}}$ unknowns per surface S_j. However, this number is drastically reduced once current and field continuity conditions are enforced by applying the junction resolution process below. Here, a junction denotes any surface edge shared by more than one quadrilateral element. Of course,

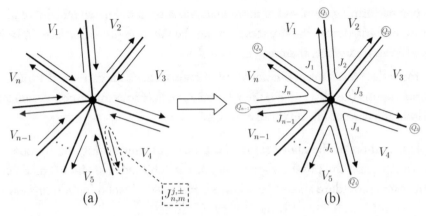

(a) $J_{n,m}^{j,\pm}$ (b)

FIGURE 4.4: 2D representation of the junction resolution process for a given n-element junction

for edges terminating the mesh (i.e., consisting of one surface element), the current quantities defined on those edges should be equivalently zero. Thus, this case of the junction resolution is assumed to be self-evident. Below, we present a four step junction resolution algorithm for "resolving" (i.e., removing current unknowns which are identical to others or refer to the same edge) currents at junctions. The simplicity of the algorithm should be greatly appreciated, though other systematic approaches have been presented in [114, 116, 117].

To begin with, let us consider the process of resolving the electric currents at the junction containing n elements as shown by its 2D representation in Fig. 4.4. Each element connected to the junction contributes two unknowns, namely J^+ and J^-, implying that there are a total of $2n$ unknowns defined at this junction. These unknowns correspond to currents that radiate into volume V_i (see Fig. 4.4). Although any continuity or boundary condition can be implemented, the three considered here are the dielectric continuity condition and the PEC and PMC boundary conditions. Specifically, the dielectric continuity condition requires that the tangential electric and magnetic fields be continuous across the interface surface. Additionally, the PEC and PMC boundary conditions require the tangential electric field and magnetic field to be zero, respectively. As a result, the pertinent junction resolution algorithm to resolve a given n-element junction consists of the following steps:

1. Pair all unknowns that radiate into the same volume, as done in Fig. 4.4(b). The resulting current unknowns are now denoted by J_i; this step reduces the total number of unknowns by half.

2. Eliminate all unknowns defined over PMC surfaces and/or radiating into an enclosed PEC surface.

3. Upon performing the above, if more than one current unknown remains, enforce dielectric continuity across the surfaces forming the junction (i.e., if surface Q_i in Fig. 4.4(b) is a dielectric surface, then set $J_{i+1} = -J_i$).

4. If more than one current unknown still remains, check for currents that radiate into the same volume (i.e., if $V_i = V_j$, then J_i and J_j radiate into the same volume) and enforce Kirchoff's current law on those currents.

After implementation of the above steps, the process of minimizing the number of current unknowns to represent the physics at the junctions is achieved. For the magnetic current unknowns, the same procedure should be followed. However, the elimination process taken over PMC surfaces will instead be performed over PEC surfaces.

4.4.2   MoM System Development

The MoM system is constructed by substituting the current and field expansions (4.34)–(4.36) into the variational form of the VSIE in (4.31)–(4.33). As discussed, the VIE in (4.31) is tested using the testing function described in (4.37), while the SIEs in (4.32) and (4.33) undergo the typical Galerkin's testing where the testing functions are equivalent to the current expansion functions. After carefully carrying out the junction resolution algorithm, which essentially provides the road map between the local and global unknowns, the matrix assembly process results in the following MoM system:

$$
\begin{bmatrix}
\bar{\mathbf{Z}}^{EE} & \bar{\mathbf{Z}}^{EM} & \bar{\mathbf{Z}}^{EJ} \\
\bar{\mathbf{Z}}^{ME} & \bar{\mathbf{Z}}^{MM} & \bar{\mathbf{Z}}^{MJ} \\
\bar{\mathbf{Z}}^{JE} & \bar{\mathbf{Z}}^{JM} & \bar{\mathbf{Z}}^{JJ}
\end{bmatrix}
\begin{bmatrix}
\mathbf{E} \\
\mathbf{M} \\
\mathbf{J}
\end{bmatrix}
=
\begin{bmatrix}
\mathbf{b}^{E} \\
\mathbf{b}^{M} \\
\mathbf{b}^{J}
\end{bmatrix}.
\tag{4.39}
$$

The matrix elements within the MoM blocks in (4.39) are filled using entries of the local matrices constructed through element-to-element testing (see Section 4.4.2). For the surface currents, this mapping is described by the junction resolution algorithm that relates the local currents defined in (4.34) and (4.35) to the global currents. Likewise, for the electric field unknowns, this mapping implies explicit enforcement of tangential field continuity across the hexahedral element faces. We remark here that the magnetic surface current unknowns could be paired with their corresponding electric field unknowns through the relationship given in (4.15), thus, reducing the total number of unknowns. However, no such pairing is enforced in this work because we wish for the PMCHWT formulation to result the in cases when only piecewise homogeneous objects are considered. In addition, as is well known for the PMCHWT formulation, the SIEs in the lower two rows in (4.39) are void of internal resonances as long as no partitioned volume region is completely enclosed by either a PEC or PMC surface. For such cases, a combined field integral equation (CFIE) as described in [68] should be used. A CFIE version of (4.39), although straightforward to derive, is not given in this work since the need only arises when closed PEC/PMC bodies are present in the solution domain.

Element-to-Element Evaluations
Submatrices $\bar{Z}^{EE}$, $\bar{Z}^{EM}$, $\bar{Z}^{EJ}$, $\bar{Z}^{ME}$, $\bar{Z}^{MM}$, $\bar{Z}^{MJ}$, $\bar{Z}^{JE}$, $\bar{Z}^{JM}$, and $\bar{Z}^{JJ}$ consist of a linear combination of values resulting from the element-to-element testing throughout the computational domain. As mentioned, this linear combination is dictated by the resolution algorithm linking the elemental (local) unknowns to the system (global) unknowns. It should be noted that testing only occurs between elements containing equivalent currents (volume or surface) that radiate within the same volume. Explicitly, if we assume that everywhere $\bar{\bar{\mu}}_r = \bar{\bar{I}}$, then the entries in the MoM submatrices consist of values found from the following expressions:

$$
Z_{mn}^{EE} = \int_{\Omega_m^V} dr\ \mathbf{h}'_m \cdot \mathbf{h}_n - k_n^2 \int_{\Omega_m^V} dr \mathbf{h}'_m \cdot \int_{\Omega_n^V} dr' g_{k_n}(\bar{\bar{\varepsilon}}_\Delta - \bar{\bar{I}}) \cdot \mathbf{h}_n
$$
$$
+ \oint_{\Omega_m^V} dr(\hat{n} \cdot \mathbf{h}'_m) \int_{\Omega_n^V} dr' \nabla' g_{k_n} \cdot (\bar{\bar{\varepsilon}}_\Delta - \bar{\bar{I}}) \cdot \mathbf{h}_n,
\tag{4.40}
$$

$$
Z_{mn}^{EM} = \int_{\Omega_m^V} dr\ \mathbf{h}'_m \cdot \int_{\Omega_n^S} dr' \nabla' g_{k_n} \times \mathbf{f}_n,
\tag{4.41}
$$

$$
Z_{mn}^{EJ} = jk_0 \mu_b^n \int_{\Omega_m^V} dr\ \mathbf{h}'_m \cdot \int_{\Omega_n^S} dr' g_{k_n} \mathbf{f}_n + \frac{j}{k_0 \varepsilon_b^n} \oint_{\Omega_m^V} dr(\hat{n} \cdot \mathbf{h}'_m) \int_{\Omega_n^S} dr' g_{k_n}(\nabla' \cdot \mathbf{f}_n),
\tag{4.42}
$$

$$
Z_{mn}^{ME} = jk_0 \varepsilon_b^n \int_{\Omega_m^S} dr\ \mathbf{f}_m \cdot \int_{\Omega_n^V} dr' \nabla' g_{k_n} \times [(\bar{\bar{\varepsilon}}_\Delta - \bar{\bar{I}}) \cdot \mathbf{h}_n],
\tag{4.43}
$$

$$
Z_{mn}^{MM} = jk_0 \varepsilon_b^n \int_{\Omega_m^S} dr\ \mathbf{f}_m \cdot \int_{\Omega_n^S} dr' g_{k_n} \mathbf{f}_n - \frac{j}{k_0 \mu_b^n} \int_{\Omega_m^S} dr\ (\nabla \cdot \mathbf{f}_m) \int_{\Omega_n^S} dr' g_{k_n}(\nabla' \cdot \mathbf{f}_n),
\tag{4.44}
$$

$$
Z_{mn}^{MJ} = -\frac{1}{2} \int_{\Omega_m^S} dr\ \mathbf{f}_m \cdot (\hat{n} \times \mathbf{f}_n) + \int_{\Omega_m^S} dr\ \mathbf{f}_m \cdot \oint_{\Omega_n^S} dr' \nabla' g_{k_n} \times \mathbf{f}_n,
\tag{4.45}
$$

$$
Z_{mn}^{JE} = -k_n^2 \int_{\Omega_m^S} dr\ \mathbf{f}_m \cdot \int_{\Omega_n^V} dr' g_{k_n}(\bar{\bar{\varepsilon}}_\Delta - \bar{\bar{I}}) \cdot \mathbf{h}_n - \int_{\Omega_m^S} dr\ (\nabla' \cdot \mathbf{f}_m) \int_{\Omega_n^V} dr' \nabla' g_{k_n} \cdot (\bar{\bar{\varepsilon}}_\Delta - \bar{\bar{I}}) \cdot \mathbf{h}_n,
\tag{4.46}
$$

$$
Z_{mn}^{JM} = -Z_{mn}^{MJ}
\tag{4.47}
$$

$$
Z_{mn}^{JJ} = \frac{\mu_b^n}{\varepsilon_b^n} Z_{mn}^{MM}.
\tag{4.48}
$$

The excitation column entries contain linear combination of the following integrals:

$$b_m^E = \int_{\Omega_m^V \cap V_0} dr \; \mathbf{h}'_m \cdot \mathbf{E}^{\mathrm{inc}}, \qquad (4.49)$$

$$b_m^M = \int_{\Omega_m^S \cap V_0} dr \; \mathbf{f}_m \cdot \tilde{\mathbf{H}}^{\mathrm{inc}}, \qquad (4.50)$$

$$b_m^J = \int_{\Omega_m^S \cap V_0} dr \; \mathbf{f}_m \cdot \mathbf{E}^{\mathrm{inc}}. \qquad (4.51)$$

The surface basis functions denoted here as $\mathbf{f}_n$ and having domain Ω_n^S are precisely those described in (4.34) and (4.35). Likewise, the volume basis functions denoted here as $\mathbf{h}_n$ and having domain Ω_n^V are those described in (4.36) with corresponding testing functions $\mathbf{h}'_n$ described in (4.37). The material parameters μ_b^n, ε_b^n, and k_n are those associated with the volume in which testing occurs.

4.5 EXAMPLES

In this section, we consider a few scattering and radiation examples that exploit some of the unique features of the proposed hybrid integral equation formulation. But first we provide a validation target to test the junction resolution algorithm.

4.5.1 Junction Resolution Validation

To validate the prescribed algorithm, we consider the test structure shown in Fig. 4.5, in which a PEC plate is partially coated with the material that actually has material parameters equal to that of free space. Thus, similar results should be obtained for the two cases when the material coating is considered and when the plate is considered by itself. It is not hard to verify that this structure will contain 1-, 2-, 3-, and 4-element junctions in the numerical implementation when a material coating is considered. This fact presents us with a good

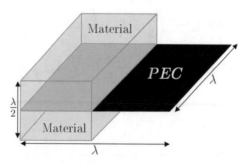

FIGURE 4.5: Illustration of a partially coated PEC plate used for validation

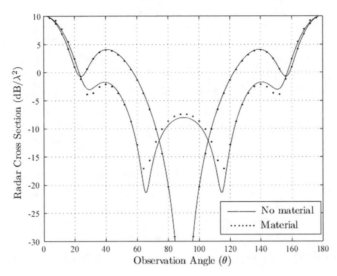

FIGURE 4.6: Bistatic (in principle plane) RCS of the PEC plate in Fig. 4.5 with the material replaced by air (for validation purposes; incident field is along $\theta = 0^0$ with θ measured from the axis normal to the plate)

benchmark to validate our junction resolution algorithm. As shown in Fig. 4.6, we obtain very similar results for both cases confirming that our integral equation implementation can handle arbitrary junctions.

4.5.2 Scattering Examples
A Homogeneous Dielectric Sphere

Consider the scattering from a modestly high contrast dielectric sphere with radius $r = 0.5\lambda_0$ and relative permittivity of $\varepsilon_r = 9$. The bistatic RCS is compared for the three different decompositions as in (4.1): (i) when $\varepsilon_r = 9 + 0 = 9(1) = \varepsilon_b \varepsilon_\Delta$, (ii) when $\varepsilon_r = 8 + 1 = 8(1.125) = \varepsilon_b \varepsilon_\Delta$, (iii) when $\varepsilon_r = 7 + 2 = 7(1.2875) = \varepsilon_b \varepsilon_\Delta$. Numerical results are compared with the analytical Mie solution in Fig. 4.7. Note here that case (i) corresponds to the PMCHWT result and as is seen, all cases produce accurate RCS solutions. Also note that a full VIE would require 19,494 unknowns with an element edge length of $0.03\lambda_0$. In contrast, the solutions considered here, which use the VIE as a perturbation, only require 9540 unknowns with an element edge length of $0.05\lambda_0$; though we note that in the limit as $\varepsilon_b = 1$ and $\varepsilon_\Delta = 9$ one should expect the solution to become more dependent on the volume field information causing the number of unknowns to increase.

Sphere Embedded in a Sphere

In our next example, we consider the case shown in Fig. 4.8. This is a high contrast dielectric sphere of radius $r = 0.1\lambda_0$ and $\varepsilon_r = 20$ embedded within a larger dielectric sphere of radius

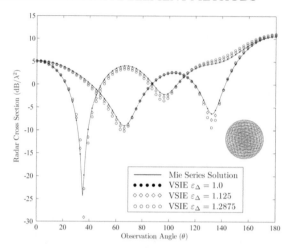

FIGURE 4.7: Bistatic RCS of a dielectric sphere of radius $r = 0.5\lambda_0$ and $\varepsilon_r = 9$

$r = 0.5\lambda_0$ and $\varepsilon_r = 10$. Our VSIE assumes an $\varepsilon_b = 10$ with $\varepsilon_\Delta = 2$ in the regions where $\varepsilon_r = 20$. To demonstrate the applicability of our method, the volume integral in (4.18) is only invoked inside the smaller embedded sphere instead of the entire scattering domain. In Fig. 4.8 we present results for two cases: i) the smaller dielectric sphere is concentric to the larger sphere; ii) the smaller dielectric sphere is translated $0.1\lambda_0$ in both the x- and y-directions. As seen, the

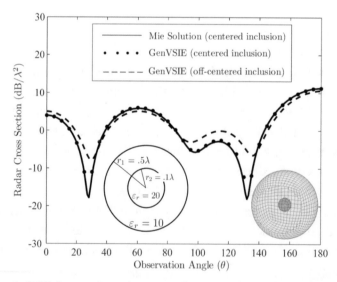

FIGURE 4.8: Bistatic RCS (θ-cut, E_θ-polarization, $\mathbf{k}_{\text{inc}} = -k_z$) of a dielectric sphere with $r = 0.1\lambda_0$ and $\varepsilon_r = 20$ embedded in a dielectric sphere with $r = 0.5\lambda_0$ and $\varepsilon_r = 10$. The dotted curve is shown for a centered inclusion (compared to Mie solution) and the dashed result is shown for an off-centered inclusion (dashed line), where a change in the scattered field is seen that demonstrates possible use of the VSIE for inverse scattering applications

first case compares well with the Mie series solution, whereas the second case demonstrates the formulations' effectiveness to model field variations among displaced inclusions (a capability well suited for inverse scattering applications). We remark that the discretization of the electric field within the VIE portion of the formulation was carried out using elements of average edge length equal to $\lambda_0/10$ instead of the typical $\lambda_0/40$ required by the contrast of the materials used in this example. As a result, we only needed 144 volume field unknowns and 5400 surface current unknowns to accurately solve the system. By comparison, if the usual VIE was to be used throughout the entire scattering domain, the dielectric contrasts would require approximately 64,000 volume elements requiring approximately 200,000 volume unknowns. We note that a similar number of unknowns would be required using the usual FE–BI formulation. Because of the much smaller number of unknowns needed for the proposed formulation, it is attractive for inverse scattering applications where one is searching for high contrast inclusions embedded within large material regions. Such problems are often encountered in biomedical microwave imaging applications [129].

Coated Ogive

We next consider a coated ogive (an elongated body of revolution), commonly used as a validation target [130]. Here, the coated ogive has a length $2\lambda_0$, a base radius of $0.24\lambda_0$, and a coating of thickness $0.07\lambda_0$, with $\varepsilon_r = 2.56 - j0.5$. We can of course choose to model the dielectric coating using the conventional VSIE or the PMCHWT-SIE method. However, to demonstrate the robustness of the VSIE formulation, we arbitrarily chose ε_Δ to be $1 + j1$ corresponding to the factorization

$$
\begin{aligned}
\varepsilon_r &= (1.03 - j1.53) + (1.53 + j1.03) \\
&= (1.03 - j1.53)(1 + j1) = \varepsilon_b \varepsilon_\Delta.
\end{aligned}
\tag{4.52}
$$

Fig. 4.9 compares results from the generalized VSIE and those from [130]. It can be seen that with the arbitrarily selected complex value for ε_Δ, our formulation compares well with the PMCHWT method and the VSIE results given in [130]. For clarity, the factorization presented in (4.52) corresponds to a homogeneous material coating with $\varepsilon_b = 1.03 - j1.53$ modulated by the variation $\varepsilon_\Delta = 1 + j1$ so as to represent a material having an $\varepsilon_r = 2.56 - j0.5$. Note also that the chosen background medium in (4.52) is lossy. This choice ensures that the solution remains numerically stable (i.e., no gain within the material), leading to a well-conditioned matrix systems.

To further demonstrate the robustness of our method, a higher contrast material is presented in Fig. 4.10 where the permittivity of the ogive coating is increased to $\varepsilon_r = 17.97 - j14.57$. This coating will be modeled using a background $\varepsilon_b = 3.00 - j10.0$ and a

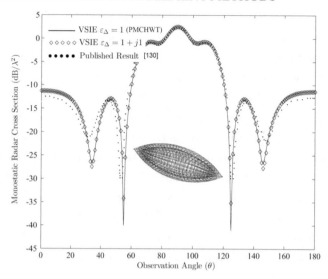

FIGURE 4.9: Monostatic RCS of a coated ogive; the coated ogive has a length $2\lambda_0$, a base radius of $0.24\lambda_0$, and a coating of thickness $0.07\lambda_0$, with $\varepsilon_r = 2.56 - j0.5$.

variation $\varepsilon_\Delta = 1.83 + j1.25$ namely,

$$\begin{aligned} \varepsilon_r &= (3.00 - j10.0) + (14.97 - j4.57) \\ &= (3.00 - j10.0)(1.83 + j1.25) = \varepsilon_b \varepsilon_\Delta. \end{aligned} \tag{4.53}$$

Again, the background medium is selected to be lossy. The results are compared to the PM-CHWT ($\varepsilon_\Delta = 1$) implementation. As shown in Fig. 4.10, we have an excellent agreement

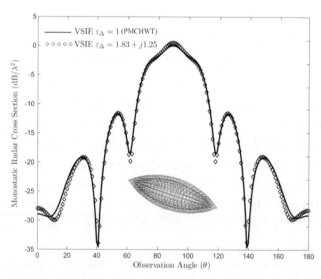

FIGURE 4.10: Monostatic RCS of the same ogive as in Fig. 4.9 with a high contrast coating having $\varepsilon_r = 17.97 - j14.57$

even down to low RCS values. It is important to further note that although the contrast of the coating was increased, the same discretization density was used for the low and high contrast examples. Since design applications often seek material profiles (i.e., ε_Δ) that optimize a desired performance parameter, the presented results suggest that this VSIE is well suited in design applications utilizing high contrast material coatings.

Smoothly Varying Inhomogeneity

In the last example, we consider the scattering from a dielectric sphere of radius $r = \lambda_0/3$ having a material profile that varies radially (see Fig. 4.11) as

$$\varepsilon_r(r) = 10 - 27r, \tag{4.54}$$

where r has units in terms of λ_0. To demonstrate the VSIE's capability to model smoothly varying material profiles, we arbitrarily select a background medium of $\varepsilon_b = 5$, for which $\varepsilon_\Delta(\mathbf{r}) = 2 - 5.4r$. In Fig. 4.12 we compare the VSIE result with that of the typical VIE solution as well as an FE–BI solution. In addition, we also incorporate a PMCHWT implementation of the inhomogeneous sphere using the layered sphere approximation with four and eight concentric layers of constant dielectric regions. As seen in Fig. 4.12, the VSIE result which required 882 VIE unknowns and 864 SIE unknowns compares well with that of the VIE result which required 1944 unknowns. Although the savings in the number of unknowns

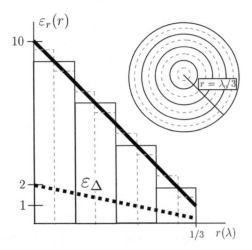

FIGURE 4.11: Material profile for the smoothly varying inhomogeneous sphere of radius $r = \lambda_0/3$. Also shown is the material profile of the perturbation, $\varepsilon_\Delta(\mathbf{r})$, for an $\varepsilon_b = 5$, along with the staircase approximation to the material profile for the four- and eight-layered sphere used within the PMCHWT implementation

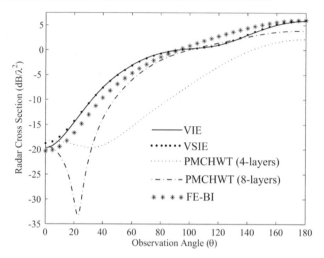

FIGURE 4.12: Bistatic RCS (θ-cut, E_θ-polarization, $\mathbf{k}_{inc} = -k_z$) of the smoothly varying inhomogeneous sphere depicted in Fig. 4.11

is not very significant, the reduction in the number of VIE unknowns drastically reduces the matrix fill time for the MoM system. Interestingly, we note that the results for both the four and eight layered spheres are substantially different from each other as well as from the VIE, VSIE, and FE–BI results. This is entirely due to the approximation error introduced in modeling the smoothly varying material profile with uniform layers. For such a staircase approximation to be valid, one would need many thin layers, which in turn could increase the computational cost of the PMCHWT formulation. For completeness, we used an FE–BI system of 6084 unknowns (with 1728 BI unknowns). Thus, because the four layered solution required 2400 unknowns and the eight-layered solution required 6000 unknowns, we can also claim that layered approximations of inhomogeneous materials may not always be the most efficient approach.

4.5.3 Antenna Examples
A Tapered Slot
To test the VSIE formulation's ability to accurately predict antenna performance, a tapered slot antenna is considered [131]. To obtain the actual dimensions of the antenna, the reader is referred to [131]. This geometry is a good example of an object that would become dominated by the boundary contributions if the FE–BI was considered. As is seen in Fig. 4.13, the calculated results agree very well with the measured data (from [131]) for both E-plane and H-plane pattern cuts.

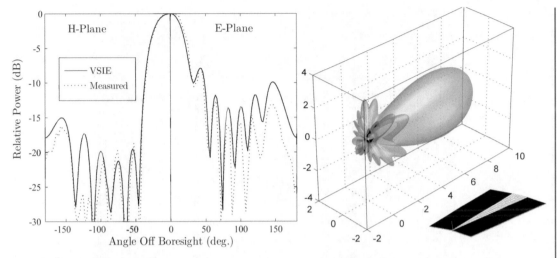

FIGURE 4.13: Tapered slot antenna—relative power and directivity patterns (measured results are after Janaswamy © IEEE, 1989 [131])

A Square-Spiral Patch

We also consider the simulation of a square-spiral patch antenna shown in Fig. 4.14 that was proposed in [132]. In order for the antenna operation band to be pushed down to very low frequencies, the patch needs to be built over a substrate having an $\varepsilon_r = 100$, a very high contrast material. We compare the results for the calculation of the real part of input impedance

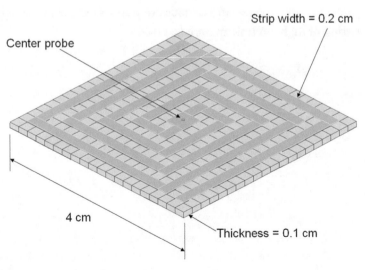

FIGURE 4.14: A square-spiral patch antenna

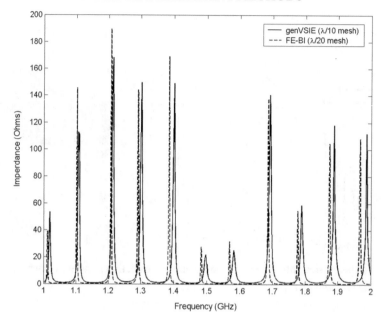

FIGURE 4.15: Real part of the input impedance of the square-spiral patch antenna, computed using the genVSIE with standard $\lambda_0/10$ mesh and the FE–BI with a $\lambda_0/20$ mesh

in Fig. 4.15 for an FE–BI-based method and the generalized VSIE method proposed here. We observed that the calculations based on the FE–BI method required a much finer mesh than did calculations based on the integral equation solution. That is, when the mesh is refined the FE–BI solution actually converged to the integral equation solution. This fact supports the consensus that integral equation based methods are well suited for modeling electromagnetic structures consisting of high contrast materials.

CHAPTER 5

Periodic Structures[1]

Periodic structures, such as frequency selective surfaces (FSSs) and volumes (FSVs), have been extensively studied and utilized in many contemporary electromagnetic systems including phased arrays, radomes, and radar absorbing materials [133–135]. The most prevalent computational tool to model such structures is the periodic moment method (PMM) discussed in [133], which utilizes a periodic Green's function inside a SIE formulation. More recently, metamaterials have commanded considerable attention due to their extraordinary electromagnetic properties [136–138]. In the previous chapters, we discussed the framework and development of several hybrid methods commonly found in frequency domain computational electromagnetics. In this chapter, we present their modifications so that one can model such periodic structures. For clarity, we choose to refer to FSSs as periodic structures consisting of only metallic surfaces, where as we refer to FSVs as periodic structures containing material volumes. Because FSVs often consist of thin material layers all periodic media are often referred to as FSS. We draw the distinction here because modeling FSSs and FSVs (especially metamaterials) inherently requires separate computational approaches.

For frequency domain simulations of periodic structures, numerical techniques based upon the hybrid finite element–boundary integral (FE–BI) method [102, 139–143] have been attractive since unit cells consisting of arbitrary material properties can be modeled by virtue of the finite element method (FEM). Recently, the hybrid volume–surface integral equation (VSIE) method presented in Chapter 4 has been extended to handle periodic media [144].

In Section 5.1, we discuss two different techniques to model periodic structures. The first is based upon the hybrid FE–BI presented in Chapter 3, and the second is based upon the hybrid VSIE presented in Chapter 4.

5.1 PERIODIC BOUNDARY CONDITIONS

In Chapters 3 and 4, we presented computational methods that could model the radiation and/or scattering from any arbitrary object. Now let us assume that this object is replicated

---

[1]This chapter is an extended version of the paper [144]

infinitely many times in two dimensions so that the objects' domains are disjoint (i.e., they do not share boundaries). The situation is depicted in Fig. 5.1. As seen, the lattice vectors $\boldsymbol{\rho}_a$ and $\boldsymbol{\rho}_b$ make up the lattice plane whose normal $\hat{\mathbf{n}}_\rho$ is given by $(\boldsymbol{\rho}_a \times \boldsymbol{\rho}_b)/A$, where $A = |\boldsymbol{\rho}_a \times \boldsymbol{\rho}_b|$.

Because both of the methods described in Chapters 3 and 4 rely upon a boundary integral formulation (or possibly a volume integral formulation in the case of the VSIE), modeling the scenario depicted in Fig. 5.1 is necessarily a simple task. Essentially, in all integral equation's involving the evaluation of the free-space Green's function $g_{k_0}(\mathbf{r}, \mathbf{r}') = \frac{e^{-jk_0|\mathbf{r}-\mathbf{r}'|}}{4\pi|\mathbf{r}-\mathbf{r}'|}$, one must instead evaluate the periodic Green's function $g_\rho(\mathbf{r}, \mathbf{r}')$ which describes the field behavior for an array of δ-sources in free space. The spatial representation of the periodic Green's function is given by

$$g_\rho(\mathbf{r}, \mathbf{r}') = \sum_{p=-\infty}^{\infty} \sum_{q=-\infty}^{\infty} e^{-j\mathbf{k}_0 \cdot \boldsymbol{\rho}_{pq}} g_{k_0}(\mathbf{r}, \mathbf{r}' + \boldsymbol{\rho}_{pq}), \tag{5.1}$$

where $\boldsymbol{\rho}_{pq} = p\boldsymbol{\rho}_a + q\boldsymbol{\rho}_b$ and

$$\mathbf{k}_0 = \pm k_0 \hat{\mathbf{n}}_\rho \times \hat{\mathbf{n}}_\rho \times \begin{bmatrix} \hat{x} \\ \hat{y} \\ \hat{z} \end{bmatrix}^T \begin{bmatrix} \sin\theta_0 \cos\phi_0 \\ \sin\theta_0 \sin\phi_0 \\ \cos\theta_0 \end{bmatrix} \tag{5.2}$$

is defined by the spherical coordinates θ_0 and ϕ_0 corresponding to the arrival angles of an incident plane wave (positive sign) or scan angles of a phased array (negative sign) [142].

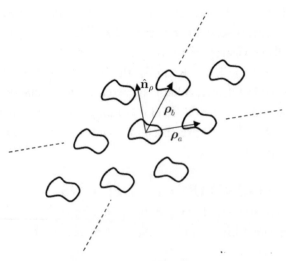

FIGURE 5.1: An arbitrary periodic structure with disjoint unit cells

At first glance, we see that the evaluation of (5.1) involves a doubly infinite summation that is actually divergent for a lossless free-space. This means that even when free-space loss is considered, accurate evaluation of (5.1) may require too many terms for practical use, especially when ρ_a and ρ_b are on the order of a few wavelengths or smaller. Instead of evaluating the spatial form of g_ρ in (5.1), it is often advantageous to consider the spectral representation given by

$$g_p(\mathbf{r}, \mathbf{r}') = \sum_{p=-\infty}^{\infty} \sum_{q=-\infty}^{\infty} \frac{e^{-j\mathbf{k}_{\rho pq}\cdot(\rho-\rho')}}{2jAk_n'} e^{-jk_n'|r_n-r_n'|}, \tag{5.3}$$

where

$$\mathbf{k}_{\rho pq} = \mathbf{k}_0 + \frac{2\pi}{A}[p(\rho_b \times \hat{\mathbf{n}}) + q(\hat{\mathbf{n}} \times \rho_a)], \tag{5.4}$$

and

$$k_n' = \sqrt{k_0^2 - \mathbf{k}_{\rho pq}\cdot \mathbf{k}_{\rho pq}}. \tag{5.5}$$

This representation, though reasonably convergent for small lattice vectors (as compared to λ_0), suffers the same fate as the spatial representation when large lattice vectors are considered. As a result of these facts, in practical implementations, most researchers evaluate g_ρ using the Ewald transform acceleration technique [145]. This approach (see [142, 146]) renders an exponentially convergent series that requires far fewer expansion terms than either the spatial or spectral representation of g_ρ.

Though the periodic scenario depicted in Fig. 5.1 can be used to model many structures of practical interest, a more ubiquitous scenario is the one in which the periodic unit cells have joining walls as seen in Fig. 5.2. In order to model such periodic structures, care must be taken when handling the fields (equivalent currents) on the walls of the unit cell, or more specifically, the periodic boundaries.

For the connected periodic structure portrayed in Fig. 5.2, the Floquet theorem must be invoked at the walls of the periodic unit cell (see Fig. 5.5). Basically, this requires that the electromagnetic quantities satisfy the periodic boundary conditions (PBCs):

$$\mathbf{E}(\mathbf{r} + \rho_{pq}) = \mathbf{E}(\mathbf{r})e^{-j\mathbf{k}_0 \cdot \rho_{pq}}, \tag{5.6}$$

$$\mathbf{M}(\mathbf{r} + \rho_{pq}) = \mathbf{M}(\mathbf{r})e^{-j\mathbf{k}_0 \cdot \rho_{pq}}, \tag{5.7}$$

$$\tilde{\mathbf{J}}(\mathbf{r} + \rho_{pq}) = \tilde{\mathbf{J}}(\mathbf{r})e^{-j\mathbf{k}_0 \cdot \rho_{pq}}. \tag{5.8}$$

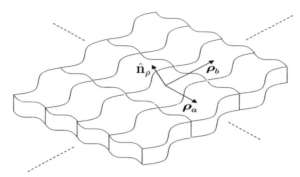

FIGURE 5.2: A connected periodic structure that requires explicit enforcement of the periodic boundary conditions [144]

In the next two sections, we will see how one must modify the two hybrid formulations from Chapters 3 and 4 so that periodic structures of the type shown in Fig. 5.2 can be properly modeled.

5.1.1   Using the FE–BI

When the connected periodic structure of Fig. 5.2 is considered within the FE–BI formalism, the three-dimensional (3D) FE–BI outlined in Chapter 3 can be simplified via the use of image theory to eliminate the unknown electric current density on the top and bottom surfaces of the domain (for the simplified FE–BI formulation applicable to domains recessed in an infinite ground plane as well as periodic domains, see [37, 102, 142, 147, 148]). Then, the remaining four side boundaries of the FE domain are joined together through the appropriate phase progressions in a much similar fashion as described above. The periodic Green's functions for the upper and lower spaces of the structure are evaluated, as before, using the Ewald transform. Even further acceleration of the iterative solution of the periodic FE–BI system was achieved in [102, 143] and was extended to include multilayered periodic Green's functions in [148].

Here we give two examples demonstrating the capabilities of the fast spectral domain algorithm (FSDA) which is based on the hybrid periodic FE–BI method mentioned above. The first example is a slot FSS structure suggested in [135]. The slot array resides on a dielectric substrate of height 0.1 cm with relative permittivity $\varepsilon = 4$. As shown in Fig. 5.3, the FSS has a band-pass characteristic with a resonance frequency of about 13 GHz, and the FE–BI calculations are in excellent agreement with the method of moments results [135]. We should mention that one additional air layer was placed on top of the original structure to move the top boundary away from the slot location. This is because if the BI surface is in the plane of the slot, the spectral distribution of the search vectors is determined by the shape of the slot. Thus,

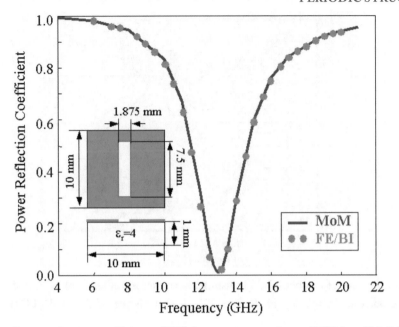

FIGURE 5.3: Power reflection coefficient of FSS slot array; comparison of FSDA and MoM results (after Eibert et al. © IEEE, 1999 [142])

especially in the direction along the width of the slot the wave spectrum is relatively broad. However, by moving the BI surface away from the slot, the spectrum of the search vectors is narrower and therefore the convergence of the corresponding Floquet mode series is faster. Thus, a fewer number of Floquet modes ensure convergence.

Our second example is a low-pass FSS structure which was investigated in [149]. This noncommensurate structure consists of eight patch layers with different (square) patch sizes and unit cell periodicities. Instead of using the original dimensions in meters, the geometry is scaled to centimeters. Similarly, the frequency must be scaled by a factor of 104. In Fig. 5.4, the transmission coefficient of the structure is shown for FE–BI calculations and these results are in good agreement with the MoM data [149].

We next proceed to demonstrate the VSIE in modeling similar FSS and FSV structures. Unlike the straightforward imposition of the PBCs in FE–BI, the VSIE PBCs require careful pairings of unknowns across the periodic boundaries of the unit cell.

5.1.2 Using the VSIE

In Chapter 4, we saw that the MoM matrix consists of a matrix assembly process that relates the local unknowns to their global counterparts. For the electric field unknowns, this process is

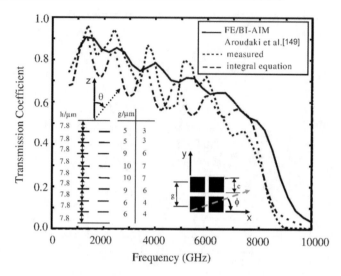

FIGURE 5.4: Transverse magnetic field transmission coefficient of an eight-layer FSS with noncommensurate periodicities ($\varepsilon_r = 2.3 - j0.08$, $\theta_0 = 30°$, $\varphi_0 = 0°$) (after Volakis et al. [147])

dictated by the satisfaction of tangential field continuity throughout the computational domain. For the current quantities, however, this process was defined by a junction resolution algorithm (see Section 4.4.1) that satisfies conservation of charge across surface element edges within the computational domain. For unconnected periodic structures, the junction resolution algorithm presented in Section 4.4.1 is pertinent. However, the standard algorithm needs to be modified in order to enforce the necessary periodicity conditions (PBCs) to model connected periodic structures as in Fig. 5.2 [144].

To properly handle the PBC for the electric field (5.6) in the VSIE formulation, all electric field unknowns defined on edges of a unit cell's walls must be paired through the proper phase term $e^{-jk_0 \cdot \rho_{pq}}$. As mentioned, the PBCs on $\tilde{J}$ and M in (5.7) and (5.8) must be satisfied using a junction resolution algorithm since the periodic boundary condition and the conservation of charge must both be explicitly satisfied. To ensure both properties, we must perform a modified junction resolution algorithm. Specifically, we can group the junctions (or element edges) on the four periodic walls of the unit cell (see Fig. 5.5) into two main groups: (1) PBC–PBC junctions and (2) aperture–PBC junctions. We define a PBC–PBC junction to be any junction (common edge) formed by exactly two surface elements defined on the unit cell's wall. If this junction lies on the vertical corner of the unit cell, there will be three corresponding PBC–PBC junctions on the other three corners; otherwise, each PBC–PBC junction will have one corresponding PBC–PBC junction on the opposite wall of the unit cell, as depicted in Fig. 5.5. Correspondingly, we define an aperture–PBC junction to be any junction consisting of one surface element defined

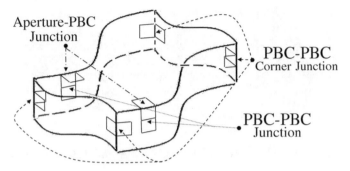

FIGURE 5.5: Types of junctions on walls of a periodic unit cell

on either the top or bottom aperture of the unit cell and one surface element defined on the unit cell's wall. For every aperture–PBC junction, there is also one corresponding aperture–PBC junction on the opposite wall of the unit cell (see Fig. 5.5).

Without loss of generality, let us consider an arbitrary PBC–PBC junction as depicted in Fig. 5.6 (where a two-dimensional (2D) representation of a junction is shown). Each arrow represents the unknown (4.34); or (4.35) corresponding to the current half-basis function defined over the surface element comprising the junction. Also, the dashed lines represent the surface elements on the unit cell's wall (PBC wall in Fig. 5.6(a)). Before junction resolution is performed, we first proceed to construct a superjunction formed by joining the two PBC–PBC junctions shown in Fig. 5.6(b). To do so, we eliminate those four current unknowns (gray arrows in Fig. 5.6(a)) that radiate into region V_0, leading to the superjunction in Fig. 5.6(b). A typical junction resolution algorithm can now be performed where any pairing of the unknowns, say by the dielectric boundary condition, across the PBC surface elements (dashed lines in Fig. 5.6(b))

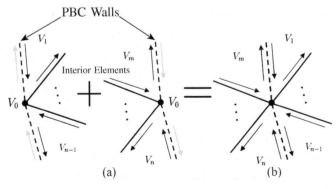

FIGURE 5.6: Two-dimensional representation of (a) PBC–PBC junctions and (b) their corresponding superjunction

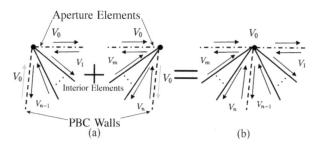

FIGURE 5.7: Two-dimensional representation of an (a) aperture–PBC junctions and (b) their corresponding superjunction

is accomplished with the appropriate phase shifting information $e^{-j\mathbf{k}_0 \cdot \rho_{pq}}$ in mind. We remark that the PBC–PBC junctions defined on the four vertical corners of a unit cell (see Fig. 5.5) are handled in the same way except that instead of two junctions, four junctions must be combined to form the superjunction.

Let us now consider an arbitrary aperture–PBC junction by referring to Fig. 5.7. Again, each arrow represents a current unknown corresponding to a half-basis function defined over a surface element, and the dashed lines represent surface elements on the periodic cell's wall. Further, the dashed-dotted lines represent a surface element on the aperture of the periodic cell. As previously done, we combine the junctions from the opposite wall to create a superjunction. This implies an elimination of the two current unknowns (gray arrows in Fig. 5.7(a)) that radiate into region V_0 and are not part of the aperture. Once the superjunction is created in Fig. 5.7(b), the junction resolution algorithm Section 4.4.1 can be invoked.

Applications

As our first example, we validate the hybrid FE–BI formulation by considering the reflection at normal incidence from an FSV studied in [142,150]. The specific FSV consists of a dielectric slab having a uniform $\varepsilon_r = 4$ with high-index blocks of $\varepsilon_r = 10$ periodically dispersed across the slab. The actual unit cell is shown in Fig. 5.8. We simulate this FSV using two different methods. In the first case, the FSV unit cell is partitioned into two homogeneous volumetric regions where the discontinuity in the material parameters is modeled using only the SIE part of the formulation (that is, the VIE is never invoked). Alternatively, we choose to formulate the FSV scattering using the VSIE where the high-index pocket is modeled using the VIE as a perturbation within the background medium of $\varepsilon_b = 4$ (with $\varepsilon_\Delta = 2.5$). As seen in Fig. 5.9, using either the SIE or VSIE portion of the formulation is identical and compares well with [142, 150]. The proposed VSIE has the flexibility to model piecewise homogeneous FSVs without a need to

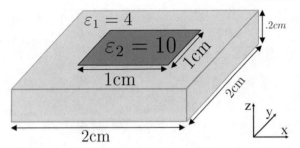

FIGURE 5.8: FSV unit cell comprised of a high-index dielectric pocket of $\varepsilon_r = 10$ within a uniform medium of $\varepsilon_r = 4$

discretize the entire volume, and this is particularly advantageous for high contrast periodic and inhomogeneous distributions within an otherwise homogeneous material region. In addition, because the costly evaluation of the periodic Green's function is unnecessary for the volume unknowns, this formulation is more efficient than standard VIEs [150]. Concurrently, the formulation retains the capability (afforded by the VIE) to model material inhomogeneities.

As the first application of the VSIE, we use it in developing an equivalent homogeneous medium using a periodic distribution of available materials [151]. For instance, to create a unidirectional metamaterial [138], one must use material layers that display large degrees of anisotropy. As an example, here we will attempt to create a material layer of thickness 1.5 cm

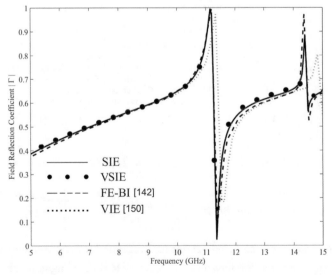

FIGURE 5.9: Reflection from FSV in Fig. 5.8 at normal incidence

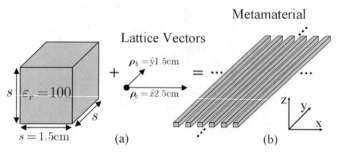

FIGURE 5.10: Anisotropic metamaterial (periodic dielectric rods) with the unit cell shown to the left

that displays the following anisotropic permittivity tensor (up to 1 GHz):

$$\bar{\bar{\varepsilon}}_{xy} = \begin{bmatrix} \varepsilon_{xx} & 0 \\ 0 & \varepsilon_{yy} \end{bmatrix} = \begin{bmatrix} 2.75 & 0 \\ 0 & 61 \end{bmatrix}. \tag{5.9}$$

One can create such a tensor by a periodic distribution of the dielectric rods depicted in Fig. 5.10. Specifically, in Fig. 5.11, we simulate the rods using our integral equation formulation with a unit cell and lattice vectors defined in Fig. 5.10(a). Unlike the typical FE–BI approaches [142], the air layers between the material rods in Fig. 5.10(b) are not included within the unit

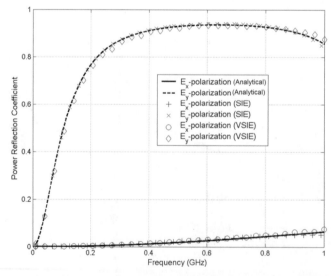

FIGURE 5.11: Power reflection coefficients at normal incidence for the periodic structure in Fig. 5.10(b) for both E_x (+) and E_y (x) polarizations; results are compared to those obtained analytically for a dielectric slab of thickness 1.5 cm having $\varepsilon_{xx} = 2.75$ and $\varepsilon_{yy} = 61$, and those obtained using the VSIE (∘ and ◇) with the perturbation described in (5.10)

cell. Further, because no anisotropies or inhomogeneities exist within the unit cell, only the equivalent electric and magnetic currents on the boundary of the unit cell need be considered in the simulation. Moreover, for $\varepsilon_r = 100$, we can expect that the VSIE formulation would require far fewer unknowns than a standard FEM approach (particularly since the air spacing between the rods is not modeled). More specifically, a standard finite element discretization of the 2.5 cm $\times$ 1.5 cm $\times$ 1.5 cm unit cell with a sampling rate at $\Delta = \frac{\lambda_0}{10\sqrt{\varepsilon_r}}$ would require $\Delta = 0.3$ cm, corresponding to approximately 225 hexahedral or 750 tetrahedral elements. In contrast, this formulation requires only 54 surface quadrilateral elements, corresponding to 168 unknowns.

To exploit the capability of the VSIE to model anisotropic materials, we now proceed to model an anisotropic layer having the material parameters in (5.9). We choose the background material of the unit cell to be $\varepsilon_b = 100$, and thus the perturbation tensor becomes

$$\bar{\bar{\varepsilon}}_\Delta = \begin{bmatrix} 0.0275 & 0 & 0 \\ 0 & 0.61 & 0 \\ 0 & 0 & 1 \end{bmatrix}. \tag{5.10}$$

As seen in Fig. 5.11, the volume equivalent sources based on the tensor in (5.10) are quite accurate in modeling anisotropies within material regions. Although the boundary integral around the unit cell results in additional surface unknowns (as compared to only using a typical VIE implementation), the matrix fill time is as much as 90% smaller. This is because the periodic Green's function is only evaluated for those surface currents which radiate outside the unit cell. One should also expect greater savings for larger unit cells.

As our next application, we consider the analysis of periodic materials consisting of resonant dielectric spheres. This example is motivated from [152], where effective medium models were constructed, and from a recent interest in using these periodic structures to emulate negative refraction index media [153–155]. For this example, we use dielectric spheres of radius 1 mm with $\varepsilon_r = 100$ that are 5 mm apart (see Fig. 5.12 for the unit cells). We will compute the transmission properties of the medium and compare the results with theoretical data based on an equivalent homogeneous medium. The latter is based on the effective medium model developed in [152]. In Fig. 5.13, we plotted the transmission coefficients corresponding to a single and dual layer of periodic spheres (solid dark plots) as compared to the theoretical solution (gray plots). We observe that the agreement between the VSIE and the effective medium data is quite good. As we noted before, the air regions between the periodic spheres are not discretized, hence, the proposed formulation is much more efficient (only 348 unknowns for the single-layer simulation and 696 unknowns for the dual-layer simulation) than the standard FE–BI implementation.

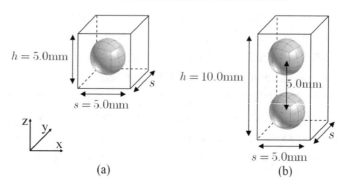

(a)   (b)

FIGURE 5.12: Unit cells for resonant dielectric spheres of radius 1 mm and $\varepsilon_r = 100$: (a) one layer of spheres, (b) two layers of spheres

Of importance in our last application is the modeling of media that exhibit left-handed material properties (effective $\epsilon_r = -1$ and $\mu_r = -1$ [153]). As seen in the inset of Fig. 5.13, the first resonant mode (TM resonance) of a dielectric sphere having radius 1.0 mm and $\varepsilon_r = 100$ occurs around 15 GHz, whereas the second mode (TE resonance) occurs around 21.2 GHz. To concurrently have $\varepsilon_r < 1$ and $\mu_r < 1$ it is necessary to include smaller spheres [137] as shown in Fig. 5.14. For the pair in Fig. 5.14, the mixing formula [154] shows (see Fig. 5.15) that the negative medium occurs at around 21.243 GHz. In Fig. 5.16, we also show the effective medium transmission properties for a slab of thickness 10 mm having the material parameters

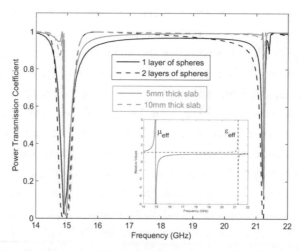

FIGURE 5.13: Calculated power transmission coefficient at normal incidence for one and two layers of dielectric spheres (radius 1 mm with $\varepsilon_r = 100$ that are 5 mm apart) compared to the theoretical solution for 5 mm and 10 mm thick slabs having material properties predicted by the effective medium models (see the inset)

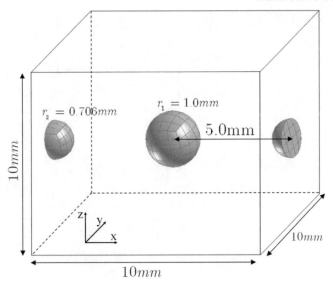

FIGURE 5.14: Unit cell consisting of two spheres (one with radius 1.0 mm and one with radius 0.706 mm) placed 5.0 mm apart in a 10.0 mm cubic lattice

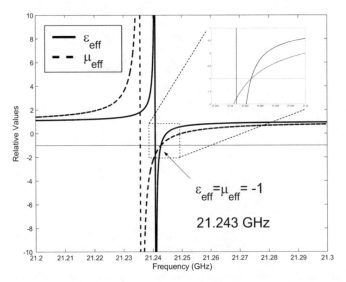

FIGURE 5.15: Theoretical μ_{eff} and ε_{eff} for a medium consisting of the spheres depicted in Fig. 5.14

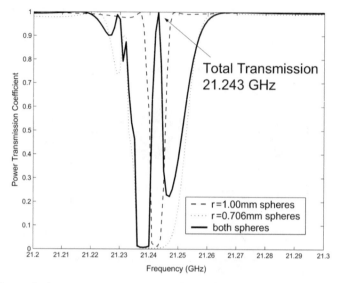

FIGURE 5.16: Theoretical transmission properties for a 10 mm thick material slab having properties equivalent to that depicted in Fig. 5.15

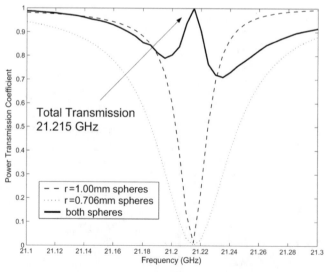

FIGURE 5.17: Calculated transmission properties at normal incidence for a medium consisting of a layer of spheres with $\varepsilon_r = 100 - j0.01$, the unit cell is depicted in Fig. 5.14

shown in Fig. 5.15. We note that at the frequency where left-handed properties appear, the slab is completely transparent. Interestingly, from Fig. 5.16 it can also be seen that if the spheres are considered in isolation, the corresponding equivalent slab would actually be opaque. This property is confirmed in Fig. 5.17 where we used VSIE to perform a full-wave simulation of the spheres depicted in Fig. 5.14. Note that the frequency (21.215 GHz) at which the left-handed material property can be expected is fairly similar to that predicted by the effective medium models (within 0.03%).

CHAPTER 6

Antenna Design and Optimization Using FE–BI Methods

by Gullu Kiziltas and Stavros Koulouridis

Design optimization has been a difficult, demanding but necessary task for the development of commercial and wireless applications, specifically for antennas and filters. Among others, these applications include conformal, multifunctional, and miniaturized antennas. Conformal antennas are greatly desired primarily because they fit easily with the curvature and contour of various geometries such as aircraft, missiles, and vehicles. Multifunctionality is critical since it eliminates the need for multiple antennas, which in turn alleviates issues of space, signal loss, and mounting.

In the past few years, we have witnessed an increasing demand for miniaturized antennas where the challenge is to maintain bandwidth and efficiency as the size is reduced. This is by definition an optimization problem solved iteratively. Historically, these problems were solved via trial and error taking months for each iteration or test. As a result, the design process relied on experience and intuition, and was often impractical. From 1960s onward, interest in automating optimal design has constantly grown. However, advanced steps in the development of numerical algorithms were prevented primarily by the lack of versatile high-speed platforms. Only in the late 1980s and 1990s field analysis tools became fast enough for integration within optimization algorithms, and thus we moved from computer-aided analysis to automated optimal design (AOD).

In electromagnetics, the essential goal of AOD is to identify a device with the desired performance and to do so in an automated manner. This is actually an inverse problem that can be solved, at least in principle, by means of numerical optimization of an objective function subject to the prescribed constraints. More specifically, the process of antenna design optimization typically consists of two modules linked together within a loop, as shown in Fig. 6.1. Starting with the initial design, the first module, i.e., the analysis module, computes the antenna performance using numerical techniques, such as integral equation (IE) and partial differential equation (PDE) methods. The second module or the synthesis module allows for the estimation

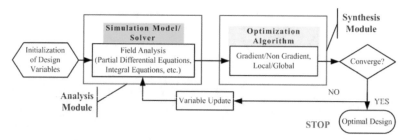

FIGURE 6.1: Formal design optimization process

of the design via an optimization algorithm. This process is repeated iteratively until some *a priori* convergence criteria are met.

The major requirements for antenna analysis and synthesis that make antenna design a challenging task and which limit the applications in antenna optimization are as follows:

Full geometrical adaptability: The simulation module must be capable of handling complex antenna geometries, such as irregular patch shapes and various feeding structures. Automatic remeshing of the geometry is also desirable during the repetitive optimization iterations, but for narrow bandwidths the same mesh could be sufficient.

Full material adaptability: The algorithms should have the capability of handling composite or artificial materials for antenna substrates and/or superstrates, resistive loading, and conducting pins (wires).

Multifunctionality: Examples of desired electromagnetic performance include maximum bandwidth, minimum size, band-pass or band-stop frequency response, prespecified scan angles, etc. Many of these objectives are in conflict with each other and tradeoffs have to be made to find a design satisfying all requirement algorithms should be able to predict various antenna performance metrics and allow for concurrent design.

Speed: Many iterations are usually required to arrive at the optimum design and most antenna problems involve a discretization scheme with hundreds of thousands of unknowns. Hence, design optimization would be unrealistic for a design unless very fast algorithms are available.

Some of the above difficulties can be alleviated by combining proper design optimization algorithms with fast and accurate solvers. Still, most existing antenna design efforts have primarily focused on *size and shape* optimization. However, recent developments in fast algorithms offer the possibility of *topology design optimization* [156] and moreover allow for full flexibility in material specification across three dimensions. To demonstrate this concept, this chapter presents a few selected optimization techniques using the hybrid finite element–boundary integral

(FE–BI) method as a solver to develop full three-dimensional (3D) antenna designs using size, shape, and topology optimization.

Our goal is to demonstrate the applicability of FE–BI methods combined with optimization techniques to develop efficient and practical antenna design problems rather than presenting an exhaustive collection of all possible techniques. For this purpose, we only give a short discussion on the most popular optimization solution techniques along with related terminology for the formal design process. Subsequently, selected techniques are demonstrated for three electromagnetic optimization design examples. Among the techniques considered, a gradient-based algorithm (sequential linear programming (SLP)) is applied to a material topology optimization problem (patch antenna design). Genetic algorithms and simulated annealing, two of the most popular gradient-free methods, are then employed and compared against each other for shape optimization of an irregular dual band patch example. Finally a multiobjective optimization via genetic algorithms is presented and applied to obtain metallo-material designs for a conformal patch antenna.

6.1   DESIGN OPTIMIZATION: OVERVIEW

6.1.1   Definition

Formally, design optimization can be defined as the effort of obtaining the "best" solution subject to prespecified design constraints. Since the effort or the desired benefit in any practical design can be expressed as a function of certain design variables, optimization can be defined as the process of finding the conditions that give the maximum or minimum value of a function $f(\mathbf{x})$. An optimization problem can be expressed in the following standard negative null form:

$$\text{Find } \mathbf{x}(x_1, x_2, \ldots, x_n) \text{ that minimizes } f(\mathbf{x}), \text{ subject to constraints } g(\mathbf{x}) \leq 0, h(\mathbf{x}) = 0, \quad (6.1)$$

where $\mathbf{x}$ is a vector with n design variables called the design vector, and $f(\mathbf{x})$ is the scalar objective function to be minimized satisfying both equality and inequality constraints $h(\mathbf{x})$ and $g(\mathbf{x})$, respectively.

6.1.2   Classification

Today, modern optimization theory offers a great variety of techniques [156, 157] for solving inverse problems in EM. To explain how knotty optimization problems are, one may try to classify them and the reader is referred to the NEOS tree proposed for a possible classification of optimization methods as depicted in Fig. 6.2. Based on the nature of the optimization models, optimization methods can also be classified into integer- or real-valued programming problems [157], single-objective or multiobjective design problems, constrained or unconstrained problems and component or system design optimization problems [156]. We focus on component optimization problems in this chapter. Traditionally, component-level design problems can be

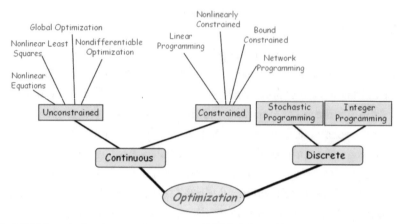

FIGURE 6.2: NEOS optimization tree (courtesy of NEOS: http://www.ece.northwestern.edu/OTC/)

classified further into size, shape, and topology optimization, where the design variables are proportions, boundaries, and topology (of the component), respectively. Here, all three types of problems are investigated.

Similar to design problems, optimization methods or the strategy/technique employed to solve the problem can be classified in various ways. Three classes of methods, however, stand out: (a) deterministic or stochastic methods; (b) local or global methods; (c) gradient-based or gradient-free methods. The determining factor is usually based on the update mechanism of the design variables as discussed below. It should be noted, however, that a specific technique does not necessarily belong to only one of the optimization groups since the classification criterion for each group is different. Hence, there are natural overlaps in the designated class.

Stochastic/Deterministic Methods

In stochastic methods some or all of the parameters (design variables and/or preassigned parameters) are probabilistic (nondeterministic). On the other hand, deterministic methods such as simplex, Rosenbrock, steepest descent, Newton–Raphson, sequential quadratic or linear programming, or Lagrangian multipliers seek the optimum point based on the information given by the gradient (sensitivity) of the objective function. Efficiency of each method depends on several factors, such as the starting point, the accuracy of the steepest descent (derivative) evaluation, the approach used to execute the line search as well as the stopping criterion. Deterministic methods usually converge to a local optimum. Two main disadvantages are the need for gradient evaluations (not always possible for real-life problems), and the lack of guaranteeing global optimum detection. Nevertheless, all techniques are mathematically well established and do not involve any heuristics ascertaining the use of minimum necessary calculation steps.

Local/Global Methods

Local methods produce results that are highly dependent on the starting point whereas global methods are largely independent of the initial guess. Also, the former tend to be tightly coupled to the solution domain. This should be rather an advantage than disadvantage since they are able to take space characteristics into account and, thus, lead to fast convergence at an optimum solution. However, the tight coupling also places constraints on the solution domain, like differentiability and/or continuity, constraints that sometimes are too difficult to be dealt with in real life. On the other hand, global methods place fewer constraints on the solution domain and are, thus, more robust when faced with an ill-behaved solution.

Gradient-Based/Gradient-Free Methods

Descent or gradient-based methods require, in addition to the objective function value, information about the first-order and possibly second-order derivatives of the given function to obtain suitable search directions toward the optimum. They guarantee descent in successive iterations and most likely fast convergence. A disadvantage of gradient-based methods is their sensitivity to the initial estimates of the design variables, although proper scaling can sometimes address this issue effectively. If the objective function has more local optima than one, the algorithm may converge to a local optimum instead of the desired global one. On the other hand, nongradient methods do not require partial derivatives of the objective function in finding the minimum. Hence, they are often called direct search methods or zeroth-order methods.

Gradient-based methods.   An early paper by Nakata and Takahashi [158] must be quoted in connection with gradient-based optimization applied to electromagnetics. Nakata and Takahashi presented a new design method for a permanent magnet using the finite element method. Subsequently, many other authors employed the same method and several variations exist in the sensitivity evaluation or the employed analysis module. Gradient-type methods have nowadays become prominent for large-scale topology optimization dealing with thousands of design variables. They have been applied to solve diverse multidisciplinary topology optimization problems, as opposed to the case of simple compliance optimization problems which could be solved using optimality criteria or intuition-based algorithms.

The gradient of a function has a very important property. If we move along the gradient direction from any point in the n-dimensional space, the function value increases at the fastest rate and so the gradient direction is called the direction of steepest ascent. Hence, the negative of the gradient vector denotes the direction of steepest descent and thus, a steepest descent method is expected to reach the optimum (minimum) faster than other techniques.

Sequential linear programming (SLP).   Similar to other gradient optimization solvers such as the sequential quadratic programming (SQP), SLP entails the sequential solution of an approximate linear subproblem. More specifically, in each optimization iteration, the objective function and constraints are replaced by linear approximations obtained from the Taylor series expansion about the current design point (design vector), **x**. The linear programming subproblem is then posed to find the optimal design change vector $\Delta\mathbf{x}$ from the current design point at each iteration k. This can be mathematically stated as

Minimize the linearized objective function, $f(x_{(k)}) + \sum_{i=1}^{N} (\Delta x_i) \left[\frac{\partial f}{\partial x_i} \right]_{x_{(k)}}$

for N design variables with the optimal design change vector being $\Delta\mathbf{x} = \mathbf{x}_{(k+1)} - \mathbf{x}_{(k)}$,

subject to the jth linearized constraint g,

$$g_j^{\min} - g_j(\mathbf{x}_{(k)}) \leq \sum_{i=1}^{N} (\Delta x_i) \left[\frac{\partial g}{\partial x_i} \right]_{x_{(k)}} \quad and \quad (\Delta x_i)_{\min} \leq (\Delta x_i) \leq (\Delta x_i)_{\max}.$$

The last set of constraints are the move limits, with $(\Delta x_i)_{\min}$ and $(\Delta x_i)_{\max}$ being the lower and upper bounds, respectively, of the allowable change in the jth design variable. This move-limit strategy is important for a stable convergence of the algorithm. After optimization of one subproblem, a new set of design variables, i.e., $\mathbf{x}_{(k+1)} = \mathbf{x}_{(k)} + \Delta\mathbf{x}$, is obtained and updated. To solve the linearized subproblem above, we need to find the sensitivities of the electromagnetic performance metrics with respect to changes in the design variables x_i. This can be done via the adjoint method as is extensively discussed in [159] for a material optimization of a patch antenna substrate subject to the maximum bandwidth performance.

Other widely used gradient-based optimizers are the sequential quadratic programming (SQP) and the generalized reduced gradient (GRG) optimizers. The SQP algorithm [160, 161] uses the second-order Taylor series expansion to approximate the true function (a local quadratic approximation) given the starting point with a linear approximation on the constraints. The SQP has been studied extensively and improved over the years [156].

Gradient-free methods.   Gradient-free optimization techniques in electromagnetic problems were introduced in the 1990s with simulated annealing (SA) and genetic algorithms (GA) being the most popular and widely used methods. Although GAs were known since 1975 [162], the first paper in the literature applying GAs to electromagnetic problems was published in 1994 [163]. Indeed, GAs, which are part of the evolutionary algorithms, seemed adequate for the solution of global EM optimization problems that usually have nonconvex and very often stiff, nondifferentiable and ill-conditioned objective functions. Simulated annealing [164] is analogous to the thermodynamic behavior of an annealed solid system slowly cooled to reach its lowest energy rate.

In addition to the GAs and SA, other methods have also been proposed. Neural networks [165] have been used for the solution of electromagnetic field problems over the past decade. Also, fuzzy programming seems useful in the solution of specific EM problems [166]. Finally, particle swarm optimization (PSO) (which belongs to the evolutionary techniques as GAs do) has been considered recently [167].

Genetic algorithms.   Many design problems are characterized by mixed continuous-discrete variables, discontinuous and nonconvex design spaces. Standard nonlinear optimization algorithms could be inefficient and computationally expensive while GAs [168] are well suited for solving such problems. GAs are based on the Darwinian concept of natural selection and evolution. The basic elements of natural genetics—survival, reproduction, and random mutations—are used to perform the search procedure. In GAs, the parameters of the design are encoded into *genes* while the combined genes form a *chromosome*, or simply a solution to the problem. Chromosomes are then gathered together to create a *population*. A new population is formed at each *generation* (iteration). Also members of the current population are called *parents* whereas the members of the subsequent population are called *children*. The merit of a chromosome or the outcome of our objective function is referred to as the *fitness*. In GAs the design variables (genes) are usually (but not necessarily) represented by strings of binary digits.

As already mentioned, three basic operations are performed during a GA cycle (or a generation). Referring to Fig. 6.3, these are selection (or survival of the best), crossover (or reproduction), and mutation. While there are several selection schemes, the most typical is the tournament procedure where a pair of chromosomes is chosen randomly from the population and the best (in terms of fitness) is selected to take part in the crossover phase. Crossover operation refers to the mating of two parents. At single point crossover (see Fig. 6.4(a)) parts of the parents' chromosomes are swapped at random locations creating two children. Crossover occurs with a certain probability p_{cross}. Among mutation schemes, of importance is the jump mutation (see Fig. 6.4(b)) where a randomly selected bit is inverted (for binary representation)

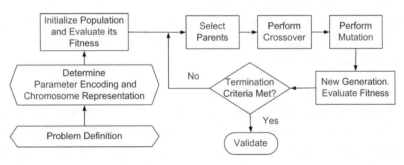

FIGURE 6.3: Genetic algorithm flowchart

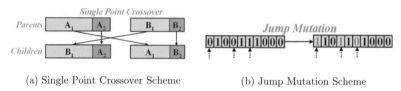

(a) Single Point Crossover Scheme   (b) Jump Mutation Scheme

FIGURE 6.4: Crossover and mutation schemes

from 1 to 0 or vice versa and this is done with the preassigned probability p_{mut}. While crossover is used to preserve the characteristics of the parents (having survived because of their good fitness), mutation is used to create new characteristics into the generation.

The GA flowchart process is shown in Fig. 6.3. After defining the problem, we "encode" the design vector into their chromosome equivalent. If the optimization model contains constraints, they can be included as penalties in the fitness function or encoded directly into the solution strings. An initial population is created and its fitness is calculated by the solver. Subsequently, we enter in the main optimization loop, where selection, crossover, and mutation operations are performed. Good solutions survive whereas poor solutions subject to the constraints are discontinued. Mutation is usually allowed with a very small probability, and is done to ensure diversity in new generations. This process is repeated until the termination criteria are met. Then, the optimal binary string is decoded back into the corresponding design. In GA, local optima are avoided by hyperplane sampling in the Hamming space (i.e., crossovers) plus random perturbations (i.e., mutations) [168].

Simulated annealing.   Simulated annealing was proposed by Kirkpatrick et al. in 1983 [169]. SA is a stochastic hill-climbing algorithm based on an analogy with the physical process of annealing. In physics of condensed matter, annealing is known as a thermal process for obtaining low-energy states of a solid in a "heat bath" [164]. The process contains two steps:

- Increase the temperature to a value at which the solid starts to melt. In this liquid state, all particles arrange themselves randomly.

- Decrease the temperature slowly until the particles take place in the ground state of the solid. In this ground state, the particles are positioned in a highly structured lattice and the system energy is minimal.

Starting from an initial vector x_0 the algorithm generates successively improved points x_1, x_2, etc., moving toward the global optimum solution. The new point $x_{i+1} = x_i + s_i$ is accepted if it satisfies the *metropolis criterion*: accept x_{i+1} if $f(x_{i+1}) - f(x_i) \leq 0$ (objective function improved) otherwise accept it with a probability $P = \exp(-\frac{f(x_{i+1}) - f(x_i)}{kT})$. Here k is a scaling factor corresponding to Boltzman's constant and T is a parameter corresponding to temperature. The value of k influences the convergence characteristics of the method leading to various cooling schedules. The simulated annealing algorithm (shown in Fig. 6.5) starts with a "high" temperature, T_0, and a sequence of design vectors is then generated until equilibrium is reached; that is, the average value of f stabilizes as i increases. During this phase, the step vector is adjusted periodically to better follow the function behavior. The best point is recorded as $\mathbf{x}_{opt}$. Once thermal equilibrium is achieved, T is reduced and a new sequence of moves is initiated starting from $\mathbf{x}_{opt}$ until a new balance point is reached. The process continues until a sufficiently low temperature is reached where no further improvements of the objective function can be expected. The essence of SA is that the optimization process is not required to go always downhill but is also allowed to make occasional uphill moves. The probabilities of accepting an uphill move and step size are determined by the value of temperature T, both of which are reduced as the temperature becomes lower. At the beginning of the SA process, T is relatively large, making the step size large and the acceptance probability high. Thus, more designs can be explored within the domain. As SA progresses, T decreases. The step size is also decreased and uphill moves are more likely to be rejected, constraining the search to a more "local" area. Eventually, the process settles by only accepting downhill moves. In this maneuver, SA prevents itself from getting stuck in inferior local optima, and is more likely to settle in areas of global quality, especially if the objective function has an overall trend toward a global optimum.

The disadvantage of SA is that parameter choices such as initial temperature, the relationship between step size and temperature, number of iterations at each temperature, and the temperature decrease rate at each cooling step are often made through trial and error. Recently, an adaptive simulated annealing (ASA) algorithm [170] was introduced to improve the basic SA algorithm. SA is similar to GA (for single-size population) when crossover is disabled and only mutations are used. SA does of course have its unique characteristics and cannot be simply seen as equivalent to GA.

Surrogate models—guiding the optimization.   Optimization problems demand calculation of the objective function several times. However, since this is a relatively time-consuming process, the

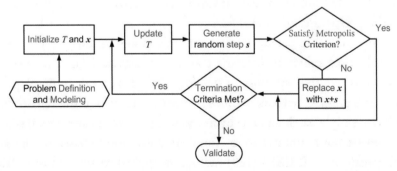

FIGURE 6.5: Simulated annealing flowchart (after Li et al. © IEEE, 2002 [178])

"best" optimization scheme is the one which requires the "least effort." In that sense, there is interest in guiding the optimization algorithm toward the "best" next solution or even replace the original complex optimization problem with a much simpler one which can be solved in considerably fewer iterations. The latter is based on surrogate models which are approximations of the objective function. Using a surrogate model provides a smooth functional approximation to the true function that is also computationally quick. This approximation can then be used in conjunction with different optimization approaches leading to a hybrid scheme.

Approximation methods have long been utilized to create surrogate models. Some of the most commonly employed methods are polynomial interpolations, artificial neural networks, and Kriging [171]. The concept of using Kriging as interpolation functions originated from mining data analysis. It is a special form of an interpolation function that employs correlation between neighboring points to determine the overall function at an arbitrary point. Kriging interpolation functions have been shown to provide good fitting in difficult problems and could lead to optimum points in a few or even tens of iterations [172]. However, it is disputable whether Kriging can be effectively employed for problems with a large number of design variables (more than 10).

Multiobjective optimization. Multiobjective optimization, also known as vector minimization, refers to a problem where more than one goal is pursued concurrently. For these kinds of problems the Pareto optimum solution is used. For a minimization problem, if we denote with f_i for $i = 1, 2, 3, \ldots$ each of the objective functions we need to optimize, a vector $\mathbf{x}^*$ is then called Pareto optimal if there does not exist another $\mathbf{x}$ in the feasible region such that $f_i(\mathbf{x}) \le f_i(\mathbf{x}^*)$ for all $i = 1, 2, 3, \ldots$ and $f_j(\mathbf{x}) \le f_j(\mathbf{x}^*)$ for at least one j [157]. This implies that no other feasible solution exists which reduces one of the objective functions without causing a simultaneous increase in at least one other objective function. This definition allows the collection of a set of optimal solutions which form the well-known Pareto front.

6.2 DESIGN EXAMPLES

6.2.1 Example 1: Dielectric Material Optimization of a Patch Antenna via Topology Optimization and SLP

As is well known, microstrip patch antennas suffer from low bandwidth. Further, their bandwidth is reduced as the substrate dielectric constant is increased for miniaturization. In this section, we demonstrate the capability of topology design methods to develop a small patch antenna subject to a prespecified bandwidth. The main goal is to improve the bandwidth performance of a chosen simple patch antenna by introducing a new (metamaterial) substrate texture whose properties are not found in nature. Topology optimization based on the solid isotropic material with penalization (SIMP) [173] method is applied to solve of the design problem.

The proposed method is aimed at designing the inhomogeneous structure of the dielectric substrate on which the patch is printed. Patch shape can also be used for further bandwidth improvements, but is not considered here.

SIMP is an attractive approach in the engineering community because of its simplicity and efficiency. It is integrated for this example with the FE–BI method [159] to remove limitations on geometry or material distribution. A key aspect of the design method is to consider design as a material distribution problem for optimization. For electromagnetic applications, these properties are the permittivity and permeability of the material as well as the conductivity or the resistance of the metallic patches. In practice, to specify the material properties, the design space is discretized into material cells/finite elements. Actually, the most straightforward image-based geometry representation is the 0/1 integer choice, where the design domain is represented by either a void or a filled/solid material (which is employed for metallization design in the second example and for concurrent conductor and material design in the third design example). However, this formulation may not be well posed mathematically [174]. To obtain a well-posed optimization problem, relaxation in the material properties is introduced using graded material properties. This is the essence of the SIMP method in which material grading is introduced using a single density variable, ρ, and relating it to the actual material property of each finite element through a continuous functional relationship. A suitable interpolation for the permittivity (could also represent resistance of the patch surface) is

$$\rho = (\varepsilon_{\text{int}} - \varepsilon_{\text{air}})/(\varepsilon_{\text{orig}} - \varepsilon_{air})^{1/n} \tag{6.2}$$

where n is a penalization factor, and ε_{int} and $\varepsilon_{\text{orig}}$ are intermediate and original solid material permittivity, respectively. Of importance is that the on/off nature of the problem is avoided through the introduction of the normalized density ρ. This parametrization also allows us to formulate the problem in a general nonlinear optimization framework. The goal is to arrive at the optimum distribution of material (densities) such that a certain performance merit is optimized subject to the design constraints.

Solution Procedure

An appropriate objective function is to maximize the return loss bandwidth sampled discretely over N_{freq} frequency points [175], namely,

$$f(\rho) = \min[\max(|S_{11}|_j)] \quad j = 1, \ldots, N_{\text{freq}} \tag{6.3}$$

subject to the volume constraint

$$\sum_{i=1}^{N_{\text{FE}}} \rho_i \cdot V_i \leq V^* \tag{6.4}$$

and side constraints

$$0 < \rho_{\min} \leq \rho_i \leq \rho_{\max} \quad i = 1, \ldots, N_{\text{FE}}. \qquad (6.5)$$

The volume constraint is imposed to limit material usage. That is, the material is confined to a maximum volume V^* within the design domain. The actual material is comprised of density ρ_i and volume V_i of each design cell in the FE domain.

The above design problem is easily recognized as a general nonlinear optimization problem with several thousand variables for a realistic 3D design composed of many design cells. This makes gradient-based optimization techniques such as sequential linear programming (SLP) favorable. The iterative optimization scheme chosen here is the sequential linear programming (SLP) method employing the DSPLP package in the SLATEC library due to its well-known efficiency and reliability. The design problem has 4000 design variables with all possible 3D combinations (using 2 materials) being 2^{4000}. Therefore, use of stochastic optimizers such as GAs or SA implies use of significant computing resources. An essential aspect of the optimization scheme is the evaluation of the EM response (bandwidth) sensitivity with respect to changes in the design variables (dielectric permittivity). Here the adjoint variable method [176] is adopted to enable versatility and fast convergence using first-order mathematical programming algorithms. It is also of great importance to impose constraints for the design changes known as move-limit bounds to ensure convergence. Here, a move-limit strategy originally proposed by Thomas et al. [177] is adopted. This strategy is known to avoid the locking of the move limits in the case of exceedingly small design variables. The basic principle is to set the value for the change in the ith design variable (say Δx_i) in accordance with the following criterion,

$$\Delta x_i = \max(Cx_i, (\Delta x_i)_{\min}) \qquad (6.6)$$

where C is a constant move ratio, and $(\Delta x_i)_{\min}$ is the minimum move limit for the ith design variable. For this example, C is set to 0.1 and $(\Delta x_i)_{\min}$ is set to approximately 1% of the upper bound of the design variable, i.e., 0.001. However, when the objective function is close to the optimal value the move ratio is decreased to half of its original value.

Design Results

The chosen geometry for optimization is a square patch excited with a probe/coax feed as displayed in Fig. 6.6(b). The frequency range of interest is 1–2 GHz sampled over 21 frequency points. Also, the volume constraint is set to 70% with respect to the initial substrate dielectric permittivity. The algorithm for the proposed design cycle follows that in Fig. 6.1. Thus, the design cycle starts with the initialization of the design variables, corresponding to an initial homogeneous dielectric substrate with a certain permittivity value. Here, for the initial design we used a homogeneous substrate of $\varepsilon_r = 42$ giving an initial response having a 6.7% return less bandwidth (calculated at -5 dB). With each design cell being updated via the SIMP method

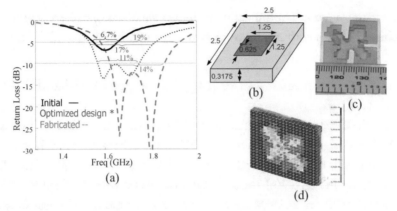

FIGURE 6.6: (a) Return loss behavior for the initial, optimized, and fabricated substrates; (b) schematic of patch antenna on the initial homogeneous substrate (dimensions are in cm); (c) fabricated two material composite substrate patch (LTCC with $\varepsilon_r = 100$ filled with epoxy stycast of $\varepsilon_r = 3$); (d) optimized material substrate with a corresponding density (ρ) scale (after Kiziltas et al. © IEEE, 2004 [159])

and the SLP routine, a heterogeneous design was obtained in only 20 iterations. Convergence was achieved when the objective function value changed less than 10^{-3} between two successive iterations. The optimized material distribution is displayed in Fig. 6.6(d) as a 3D color-coded block with each color pixel corresponding to a certain density/permittivity value. The corresponding return loss behavior of the optimized dielectric distribution is depicted in Fig. 6.6(a) and is compared to the initial performance.

The computation time for the entire design process with a design domain of 4000 finite elements/design cells and 21 frequency points for each iteration was 17 h on a Pentium 3 processor. Given the poor bandwidth at the starting design, the attained bandwidth performance (with material design only) is truly remarkable. Further, bandwidth improvements are possible via patch shape design. To fabricate the design, adaptive image processing with a simple filtering strategy based on a cutoff value of 0.64 for the densities was adapted. To attain a smooth surface, the light colored regions (see Fig. 6.6(d)) are filled with an epoxy stycast having a dielectric constant of 3 as depicted by the transparent material in Fig. 6.6(c). The return loss behavior of the fabricated design is compared to the initial substrate and the optimized volumetric material design in Fig. 6.6(a). We note that the bandwidth of the final fabricated design presents a threefold improvement over the initial design. To validate the performance of the metamaterial substrate, a square 1.25 cm × 1.25 cm metallic patch was formed using ECCOCOAT C-110-5 silver paint. A coaxial probe feed was then used to excite the patch. As shown, the measured bandwidth and nulls of the fabricated metamaterial substrate agree very well with the simulations except for a frequency shift of 100 MHz [159]. This small shift is attributed to the feed location and exact patch fabrication uncertainties. Above all, this example demonstrates the power of integrating robust optimization techniques with simple filtering processes to obtain manufacturable substrates with improved performance within practical CPU times.

6.2.2 Example 2: Optimization of an Irregular-shaped Dual-band Patch Antenna via SA and GA

In this example, the design of an irregular-shaped dual-band patch antenna is considered using GAs and SA. The design is carried out by combining a schematic optimizer with an FE–BI solver [178]. The goal is to design a patch that operates at two GPS frequencies, 1227 MHz and 1572 MHz. As shown in Fig. 6.7(a), the design domain has a size of 5.36 cm × 7.02 cm, residing on top of a cavity 10.72 cm × 14.04 cm and above a substrate of thickness 0.48 cm having a dielectric constant $\varepsilon_r = 4.4$. This region is discretized using 6 × 6 rectangular elements/cells. Each cell or variable is allowed to be either filled ($x_i = 1$) or empty ($x_i = 0$), although the four elements at the middle-right part of the domain are always filled with metal to fix the feed point. Also, due to symmetry requirements, only half of the domain need be considered. Therefore, we have 16 design variables $x_1, x_2, \ldots, x_{16}$. The objective function of this problem is chosen to be

$$\{|S_{11}|_1 + 0.1|S_{11}|_2 + |S_{11}|_3\}$$

where $|S_{11}|_k$, for $k = 1, 2, 3$, refers to the return loss at three sample frequency points (1227 MHz, 1400 MHz and 1573 MHz in this case). Here, more weight is placed on $|S_{11}|_1$ and $|S_{11}|_3$, with less weight placed on $|S_{11}|_2$ so that the design is pushed to have dual resonance. This

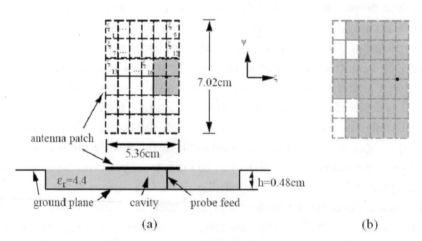

(a)   (b)

FIGURE 6.7: (a) Top and side view of the dual-frequency design domain and (b) optimal design of the patch for dual-frequency operation (after Li et al. © IEEE, 2002 [178])

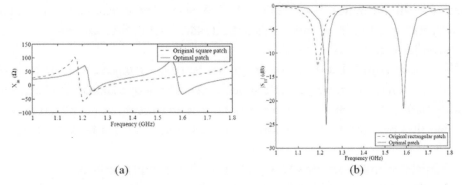

(a)   (b)

FIGURE 6.8: (a) Input impedance and impedance bandwidth of the original and (b) optimal dual-frequency patches (after Li et al. © IEEE, 2002 [178])

objective function can also be combined with other criteria based on polarization and pattern requirements. However, here the focus is only on the optimization of the return loss.

For comparison purposes, both GA and SA optimizers are separately applied to this design problem. Although not required, both methods generate the same optimal configuration shown in Fig. 6.7(b), whereas the corresponding impedance and return loss are shown in Fig. 6.8. Per design, the resonant frequencies of the optimized patch occur at 1225 MHz and 1571 MHz. However, the iteration histories for the GA and SA runs are quite different. Although GA and SA obtained the same optimal design, Fig. 6.9(a) shows that the GA algorithm finished at the 460th simulation (i.e., at the 92nd generation) whereas the SA took 850 simulations (Fig. 6.9(b)) to reach the same result. That is, the GA algorithm converged in nearly half the number of iterations (and CPU time) as compared to the SA algorithm. This might be caused by the selection of SA parameters such as step size, temperature decreasing rate, and the number of

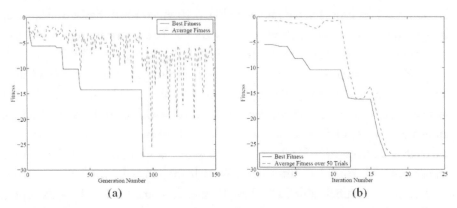

(a)   (b)

FIGURE 6.9: Iteration history of (a) GA and (b) SA for the dual-frequency patch design (after Li et al. © IEEE, 2002 [178])

iterations at each temperature. If a different set of SA parameters are chosen to run this problem, the convergence speed might change. This is the disadvantage of SA, i.e., heavy reliance on the cooling schedule parameters.

We close by noting that the above design was based on an objective function which dealt with frequency tuning. However, for this application a feed design should also be included for circularly polarized radiation. The latter was not discussed since it was not placed as part of the design objectives but can be regarded either as a concurrent or a second step optimization.

6.2.3   Multiobjective Antenna Design Using Volumetric Material Optimization and Genetic Algorithms

Multiobjective optimization is an important tool in the design of miniaturized antennas since conflicting goals such as high gain, large bandwidth, and size reduction must be addressed simultaneously. In this section, full (3D) conformal patch antenna designs are pursued using concurrent size, metallization shape as well as volumetric dielectric and magnetic material optimization. This is done by integrating genetic algorithms (GAs) and a rigorous, well-verified finite element–boundary integral (FE–BI) code [103]. A wide-frequency sweep is employed, using a single geometry model, to enhance speed, along with several discrete material choices for realizable optimized designs. Furthermore, more than one optimal design in terms of gain, bandwidth, and size is identified which may lie on several Pareto fronts. Hence, this work can be regarded as an attempt to generate several Pareto fronts or a pool of data that lie on a surface Pareto front. While there is no absolute Pareto front guaranteeing a perfect solution, the Pareto front provides the antenna designer with optimal alternatives. For example, we may collect data that lie on the Pareto curve subject to gain A_1 and bandwidth A_2, as well as data corresponding to gain B_1 and bandwidth B_2. Nonetheless, the ability to concurrently pursue magnetic, dielectric, and metallic optimization presents us with a wide range of designs [179].

The optimization approach is depicted in Fig. 6.10. After the initialization and calculation of the first generation, three antenna parameters are tabulated, namely return loss bandwidth (with strict reference to $-10\,dB$), center frequency, and center gain. The frequency sweep is carried out so that the antenna's electrical size varies within the required values while ensuring that the mesh accuracy remains acceptable at higher frequencies. The parameter of antenna gain is used to select the best candidates for each size and bandwidth. In the end, a categorization of the candidates is generated based on their gain, bandwidth, and electrical size.

As already noted, instead of exclusively looking for a single optimum design with explicitly defined constraints, we consider a multiobjective scheme. That is, we do not concentrate on the best design for an explicitly defined A or B case. In contrast, we allow the optimizer to search in a huge design space where multiple solutions can be found, each having uniquely exploitable qualities. This approach comes in contrast with known design approaches where

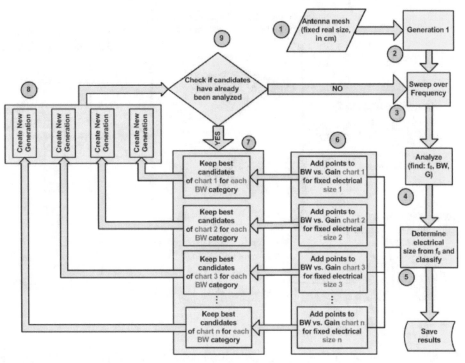

FIGURE 6.10: General multifrequency antenna optimization using a genetic algorithm (modified after Koulouridis et al. © IEEE, 2007 [179])

the engineer applies the optimization algorithm to a specific design. One could say that the design is fitted to the optimization procedure rather than the optimization being focused on a single constrained problem. Further, even if only gain is part of the objective function, the merit of a design is based on all three criteria (i.e., gain, bandwidth, frequency). Specifically, for all similar sized cases, results are accumulated with optimum gain performance for different bandwidths, allowing us to choose the best design for a particular application. Moreover, by performing antenna optimization in a vast frequency window (over 3 to 1 size) with a single mesh geometry, we significantly decrease the necessary time to investigate multiple electrical size designs. One can fit this approach into a three-goal Pareto optimization. Most often the developed antenna optimization used 2D Pareto- front except for very few cases where the 3D Pareto front was explored (e.g., in [180]). However, here the Pareto front is used to connect together three criteria based on frequency (size), bandwidth, and realized gain.

Fig. 6.11 shows the flow diagram for the generation and collection of the best/optimum solutions. The implied optimization example is a printed antenna on a textured substrate (consisting of magnetic and dielectric materials) with quad symmetry (for possible CP applications). Thus, all unique chromosomes are confined within a single quadrant. Fig. 6.11 also depicts

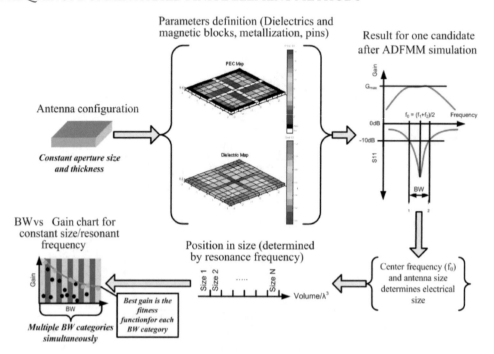

FIGURE 6.11: Illustration of how a candidate (chromosome) is examined and its fitness evaluated based on antenna gain. Subsequently, it is categorized according to bandwidth and resonance frequency (electrical size) (after Koulouridis et al. © IEEE, 2007 [179])

encoding examples for the PEC and material layers. The top color display refers to the metallization pattern on the material substrate whose distribution is shown underneath. A third patterning possibility could be the specification of loads or vertical conducting pins at each node of the grid. We note that the metallization pattern contains three groups of elements. The black elements refer to fixed metallization connecting to the feed, the white pixels refer to regions kept unmetallized during the optimization process whereas the color (or gray scale) pixels are those whose metallization will be defined via optimization.

Based on the above, each chromosome contains two parts: one refers to the metallization and the other to the material coding. The metallization portion of the chromosome is a series of bits representing the surface region with their value (1 or 0) implying metallic or nonmetallic areas. The material portion of the chromosome is defined using bit groups. Each bit group (assigned to a location) represents the chosen material of that volume element. For instance, when the optimizer has the choice of four materials (e.g., $\varepsilon_r = 1, 2.2, 4, 10.2$) two binary bits are needed to depict all material possibilities. Once the grids are created and coded, the solver evaluates the gain, bandwidth, and resonance frequency (the most time-consuming step) as

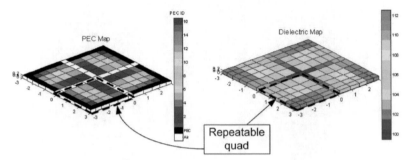

FIGURE 6.12: Metallization and dielectric volume color-coding representation (size in cm) used for the optimization of the metallo-dielectric (ε_r only) rectangular volume. Left: metallic surface of patch; right: dielectric substrate map (after Koulouridis et al. © IEEE, 2007 [179])

part of the process displayed in Fig. 6.11. Each run is then classified according to resonance frequency and sorted with respect to bandwidth whereas gain determines its fitness.

CP Designs with Dielectric Volume and Metallic Surface Optimization

A metallo-dielectric design is considered here. Fig. 6.12 depicts the quad patch geometry used for dielectric volume and metallic surface optimization. Apart from the ground plane that is kept intact, the four black L-shaped PEC strips to the left of Fig. 6.12 are also kept unchanged (this is done to induce resonances). The size of the geometry is 6 cm × 0.2 cm with 1.5–3.5 GHz for the frequency range of interest, making the electrical thickness range 0.01–0.023λ and the aperture size range 0.3–0.7λ. The permittivity values available to the optimizer were $\varepsilon_r = 1, 2.2, 6.2, 10.2, 20, 25, 30, 40$. Based on the number of dielectric and metallic choices, 50 chromosomes per population with 50 generations per size category were used to provide ample sampling of the design space. The mutation probability was set to $p_{mut} = 0.01$ (i.e., mutation performed at a probability lower than p_{mut}) and the crossover probability was set to $p_{cross} = 0.5$ (crossover occurred with a probability lower than p_{cross}). We note that a single point crossover scheme was used.

The resulting solutions from the optimization runs are given in Fig. 6.13. This is a colored 3D chart (the two dimensions are gain—vertical axis—and bandwidth—horizontal axis—while the third dimension is the electric aperture size represented by color). We should mention that 33.5% of the examined cases (chromosomes) had nonzero solutions (exhibited at least one resonance point with return loss lower than -10 dB). The performance of the two cases identified in Fig. 6.13 (labeled as cases 1 and 2) is displayed in Fig. 6.14 (note that in this figure the color (or gray scale) refers to dielectric constants whereas the axes give the location of the volume elements). These two cases represent the two extremes of the tradeoff between gain and bandwidth for approximately the same antenna size (both are 0.35λ). Specifically, case 1

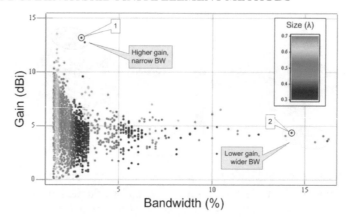

FIGURE 6.13: Gain versus bandwidth and electrical size of the optimized designs using dielectric materials and metallization (after Koulouridis et al. © IEEE, 2007 [179])

has very narrow bandwidth and high gain (>13 dBi) whereas case 2 has larger bandwidth at the expense of gain (drops down to 4 dBi, namely, a 9 dB reduction for a fourfold increase in bandwidth). However, both cases refer to very thin substrates only 0.011λ thick. Further, the patch is on a small ground plane implying a further improvement when an infinite ground plane is used.

LP Designs Based on Magnetodielectric Volume and Metallic Surface Optimizations

We now consider magnetodielectric loadings to design a miniature linearly polarized (LP) antenna. Consequently, we can choose metallic strips within and on the surface of the material volume. These are placed in a manner that minimizes cross polarization. The PEC and material

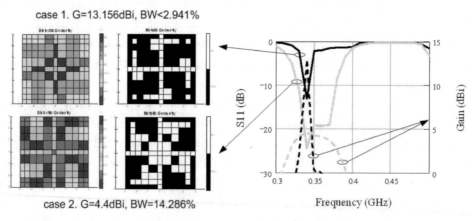

FIGURE 6.14: Optimized metallo-dielectric CP designs. Left: resulting dielectric distribution and metallic geometries; right: corresponding return loss and gain (after Koulouridis et al. © IEEE, 2007 [179])

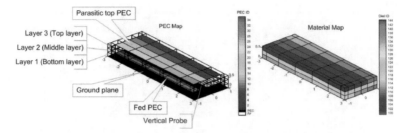

FIGURE 6.15: Linear polarized design used for magnetodielectric and metallization optimization (all sizes in cm). PEC and dielectric maps that describe the optimization regions are also shown (after Koulouridis et al. © IEEE, 2007 [179])

map geometry are shown in Fig. 6.15. As seen, the geometry consists of three material layers with PEC patterning at the top of the bottom layer and at the top of the third (uppermost) layer. A fixed metallic strip (not optimized) is created at the bottom layer. The antenna feeding is placed at a fixed point near the edge of this strip (see Fig. 6.15(a)), making the metallization on the third layer parasitic. Seven metallic strips are optimized creating 35 PEC pixels (i.e., 7 strips with 5 pixels for each strip). The number of material pixels is 45. Further, the overall size of the antenna is set to 6 cm × 2 cm × 0.6 cm with the frequency ranging from 0.5 GHz to 1.5 GHz, making the largest electrical dimension of the strip geometry 0.1λ to 0.3λ long. We specifically chose eight values for ε_r and eight values for μ_r. They are $\varepsilon_r = 1, 2.2, 5, 10, 20, 25, 30, 40$ and $\mu_r = 1, 2, 5, 8, 10, 13, 16, 20$ (not necessarily available commercially). Thus, 64 different material combinations were made available to GA. For this optimization trial, we used 100 chromosomes per generation (as the chromosomes were now obviously longer) with 100 generations per electrical size. We also increased the crossover probability to $p_{cross} = 0.7$ (as compared to $p_{cross} = 0.5$ for the CP design) to increase the search sample in the design space. However, the mutation probability was kept to $p_{mut} = 0.01$.

The resulting solutions from the optimization runs are given in Fig. 6.16. This time the "hit" ratio was 5.75%. The much smaller ratio (as compared to 33.5% for the CP patch presented above) depicts the much more difficult optimization in a design space with much fewer solutions (since it is more difficult to design antennas having sizes smaller than 0.3λ). Two of the optimized cases in Fig. 6.16 are displayed in Fig. 6.17. As seen, for the "best bandwidth", the resulting gain is remarkably constant at 4 dBi with 7.1% bandwidth for a total length of 0.27λ (including the ground plane). The second case refers to an antenna having 0.11λ for its largest length (Figs. 6.17(a) and (b) and exhibiting 4.7 dBi gain with 3.44% bandwidth.

6.3   COMMENTS

With the growing use of RF, millimeter- and microwave devices, improvements and size reduction are of interest. Concurrently, cost, size, and performance expectations become more and

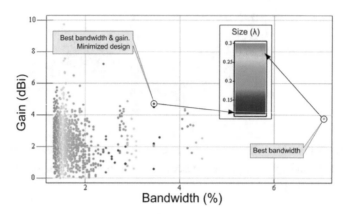

FIGURE 6.16: Gain versus bandwidth and size graph depicting the optimization results using magnetodielectric materials and metallization for the geometry in Fig. 6.9 (after Koulouridis et al. © IEEE, 2007 [179])

more stringent, necessitating versatile yet efficient design optimization procedures. In this chapter, we presented some common optimization methods to demonstrate that practical designs with improved performance are realizable. Naturally, a multitude of other optimization methods have been used in the past for specific EM problems. Here we deal with the sequential linear programming (SLP), a gradient-based method, and two gradient free methods, genetic algorithms

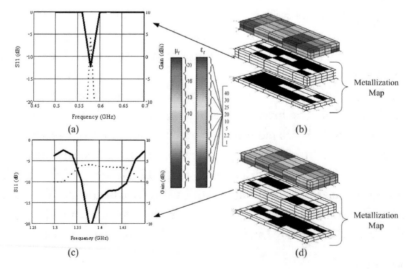

FIGURE 6.17: Examples of optimized metallomagnetodielectric LP designs. Right: resulting dielectric and metallic geometries; left: corresponding return loss and gain (after Koulouridis et al. © IEEE, 2007 [179])

(GAs) and simulated annealing (SA). We integrated these with finite element–boundary integral (FE–BI) solvers to demonstrate full flexibility in designing antennas by combining size, shape, topology optimization of conductors, and dielectric materials. Of importance is the use of fast $O(N)$ or $O(N \log N)$ algorithms for carrying out the hundreds and possibly thousands of solver calls within the design loop. We demonstrated the practicality of the integration for four antenna applications. We may also add that the possibility of material use holds great promise because it dramatically increases the available design's degrees of freedom. We can also note that topology optimization methods are more general than size and shape optimization and permit determination of the topological configuration of devices in terms of geometry, physical dimensions, connectivity of boundaries, and material implants. As we saw in the first example, a relaxed version of topology optimization allows for a mathematically well-posed electromagnetic optimization model even if it "requires" sensitivity analysis. The multiobjective optimization in the third example provides the designer with a tool to satisfy contradicting goals and let him/her choose the best solution.

As a final remark, we note that design optimization is a useful, or even necessary, tool for designing antennas. One could state that design optimization is likely to have a tremendous impact on microwave research and in overcoming contradictions to make novel RF devices.

References

[1] J. H. Richmond, "Scattering by a dielectric cylinder of arbitrary cross section shape," *IEEE Trans. Antennas Propag.*, Vol. 13, No. 3, pp. 334–341, 1965. doi:10.1109/TAP.1965.1138427

[2] R. F. Harrington, *Field Computations by Moment Methods*. New York: Macmillan, 1968.

[3] P. Silvester, "Finite element solution of homogeneous waveguide problems," *Alta Frequenza*, Vol. 38, pp. 313–317, 1969.

[4] P. L. Arlett, A. K. Bahrani and O. C. Zienkiewicz, "Application of finite elements to the solution of helmholtz's equation," *Proc. IEEE*, Vol. 115, pp. 1762–1766, 1968.

[5] P. P. Silvester and R. L. Ferrari, *Finite Element for Electrical Engineers*, 1st edition. New York: Cambridge University Press, 1983.

[6] R. G. Kouyoumjian, *The Geometrical Theory of Diffraction and its Application*, R. Mittra, Ed. New York: Springer-Verlag, 1975.

[7] S. M. Rao, D. R. Wilton and A. W. Glisson, "Electromagnetic scattering by surfaces of arbitrary shape," *IEEE Trans. Antennas Propag.*, Vol. 30, No. 3, pp. 409–418, May 1982. doi:10.1109/TAP.1982.1142818

[8] A. Taflove, K. R. Umashankar and T. G. Jurgens, "Validation of fdtd modeling of the radar cross-section of three-dimensional structures spanning up to nine wavelengths," *IEEE Trans. Antennas Propag.*, Vol. 33, No. 6 pp. 662–666, June 1985.

[9] A. Taflove and S. Hagness, *Computational Electrodynamics: The Finite-Difference Time-Domain Method*, 3rd edition. Norwood, MA: Artech House, 2005.

[10] B. Engquist and A. Majda, "Absorbing boundary conditions for the numerical simulation of waves," *Math. Comput.*, Vol. 31, pp. 629–651, 1977. doi:10.2307/2005997

[11] A. Bayliss and E. Turkel, "Radiation boundary conditions for wave-like equations," *Commun Pure Appl. Math.*, Vol. 33, pp. 707–725, 1980.

[12] W. D. Burnside, C. L. Yu and R. J. Marhefka, "A technique to combine the geometrical theory of diffraction and the moment method," *IEEE Trans. Antennas Propag.*, Vol. 23, No. 4, pp. 551–558, July 1975. doi:org/10.1109/TAP.1975.1141117

[13] G. A. Thiele and T. H. Newhouse, "A hybrid technique for combining moment methods with the geometrical theory of diffraction," *IEEE Trans. Antennas Propag.*, Vol. 23, No. 1, pp. 62–69, Jan. 1975. doi:10.1109/TAP.1975.1141004

[14] P. Monk, *Finite Element Methods for Maxwell's Equations*. Oxford University Press, 2003.

[15] A. Bossavit, *Computational Electromagnetism.* San Diego, CA: Academic, 1998.

[16] J. Nedelec, "Mixed finite elements in $\mathbb{R}^3$," *Numer. Math.*, Vol. 35, pp. 315–341, 1980. doi:10.1007/BF01396415

[17] H. Whitney, *Geometric Integration Theory.* Princeton, NJ: Princeton University Press, 1957.

[18] A. Bossavit, *Mixed Finite Elements and the Complex of Whitney Forms.* Academic Press, London, 1988, pp. 137–144.

[19] K. A. Ross, *Elementary Analysis: The Theory of Calculus.* Springer-Verlag New York, 1980.

[20] P. G. Ciarlet, *The Finite Element Method for Elliptic Problems* (Studies in Mathematics and its Applications, Vol. 4). Elservier North Holland Inc., New York, 1978.

[21] Z. Chen and Y. Xu, "The Petrov–Galerkin and iterated Petrov–Galerkin methods for second-kind integral equations," *SIAM J. Numer. Anal.*, Vol. 35, No. 1, pp. 406–434, Feb. 1998. doi:10.1137/S0036142996297217

[22] M. N. Vouvakis, S.-C. Lee, K. Zhao and J. F. Lee, "A symmetric FEM-IE formulation with a single-level IE-QR algorithm for solving electromagnetic radiation and scattering problems," *IEEE Trans. Antennas Propag.*, Vol. 52, No. 11, pp. 3060–3070, Nov. 2004. doi:10.1109/TAP.2004.837525

[23] S. N. Atluri, *The Meshless Method (MLPG) for Domain and BIE Discretizations.* Tech Science Press, Encino, CA, 2004.

[24] B. C. Usner, K. Sertel and J. L. Volakis, "Conformal Galerkin testing for VIE using parametric geometry," *Electron. Lett.*, Vol. 40, No. 15, pp. 926–927, July 2004. doi:10.1049/el:20045004

[25] V. Girault and P. A. Raviart, *Finite Element methods for Navier–Stokes Equations.* Springer, New York, 1986.

[26] P. Grisvard, *Elliptic Problems in Nonsmooth Domains.* Pitman Boston-London-Melbourne, 1985.

[27] W. McLean, *Strongly Elliptic Systems and Boundary Integral Equations.* Cambridge: Cambridge University Press, UK, 2000.

[28] R. F. Harrington, *Time-Harmonic Electromagnetic Fields.* New York: McGraw-Hill, 1961.

[29] C. Balanis, *Advanced Engineering Electromagnetics.* New York: Wiley, 1989.

[30] W. Rudin, *Principles of Mathematical Analysis*, 3rd edition. New York: McGraw-Hill, 1976.

[31] C. W. Crowley, P. P. Silvester and H. Hurwitz, "Covariant projection elements for 3D vector field problems," *IEEE Trans. Magn.*, Vol. 24, No. 1, pp. 397–400, Jan. 1988. doi:10.1109/20.43940

[32] R. D. Graglia, "The use of parametric elements in the moment method solution of static and dynamic volume integral equations," *IEEE Trans. Antennas Propag.*, Vol. 36, No. 5, pp. 636–646, May 1988. doi:10.1109/8.192140

[33] R. D. Graglia, P. L. E. Uslenghi and R. S. Zich, "Moment method with isoparametric elements for three-dimensional anisotropic scatterers," *Proc. IEE*, Vol. 77, No. 5, pp. 750–760, May 1989. doi:10.1109/5.32065

[34] M. I. Sancer, R. L. McClary and K. J. Glover, "Electromagnetic computation using parametric geometry," *Electromagnetics*, Vol. 10, pp. 85–103, 1990.

[35] S. Wandzura, "Electric current basis functions for curved surfaces," *Electromagnetics*, Vol. 12, pp. 77–91, 1992.

[36] G. E. Antilla and N. G. Alexopoulos, "Scattering from complex three-dimensional geometries by a curvilinear hybrid finite-element-integral equation approach," *J. Opt. Soc. Am. A*, Vol. 11, No. 4, pp. 1445–1457, April 1994.

[37] J. L. Volakis, A. Chatterjee and L. C. Kempel, *Finite Element Methods for Electromagnetics: Antennas, Microwave Circuits, and Scattering Applications*. New York: IEEE, 1998.

[38] L. S. Andersen and J. L. Volakis, "Accurate and efficient simulation of antennas using hierarchical mixed-order tangential vector finite elements for tetrahedra," *IEEE Trans. Antennas Propag.*, Vol. 47, No. 8, pp. 1240–1243, Aug. 1999. doi:10.1109/8.791938

[39] R. D. Graglia, D. R. Wilton and A. F. Peterson, "High order interpolatory vector bases on prism elements," *IEEE Trans. Antennas Propag.*, Vol. 46, No. 3, pp. 442–450, Mar. 1998. doi:10.1109/8.662664

[40] Z. J. Cendes, "Vector finite elements for electromagnetic field computation," *IEEE Trans. Magn.*, Vol. 27, No. 5, pp. 3958–3966, Sep. 1991. doi:10.1109/20.104970

[41] W. Schroeder and I. Wolff, "The origin of spurious modes in numerical solutions of electromagnetic field eigenvalue problems," *IEEE Trans. Microw. Theory Tech.*, Vol. 42, No. 4, pp. 644–653, April 1994. doi:10.1109/22.285071

[42] L. S. Andersen and J. L. Volakis, "Development and application of a novel class of hierarchical tangential vector finite elements for electromagnetics," *IEEE Trans. Antennas Propag.*, Vol. 47, No. 1, pp. 112–120, Jan. 1999. doi:10.1109/8.753001

[43] J. F. Lee, "Analysis of passive microwave devices by using 3 dimensional tangential vector finite element fields," *Int. J. Numer. Model., Electron. Netw. Devices Fields*, Vol. 3, pp. 235–246, 1990. doi:10.1002/jnm.1660030404

[44] J. F. Lee, D. K. Sun and Z. J. Cendes, "Full-wave analysis of dielectric waveguides using tangential vector finite elements," *IEEE Trans. Microw. Theory Tech.*, Vol. 39, No. 8, pp. 1262–1271, Aug. 1991. doi:10.1109/22.85399

[45] P. P. Silverster and M. S. Hsieh, "Finite-element solution of two-dimensional exterior field problems," *IEE Proc. H*, Vol. 118, pp. 1743–1747, Dec. 1971.

[46] B. H. McDonald and A. Wexler, "Finite-element solution of unbounded field problems," *IEEE Trans. Microw. Theory Tech.*, Vol. 20, No. 12, pp. 841–847, Dec. 1972. doi:10.1109/TMTT.1972.1127895

[47] J. D. Collins, J. L. Volakis and J.M. Jin, "A combined finite element-boundary integral formulation for solution of two-dimensional scattering problems via CGFFT," *IEEE Trans. Antennas Propag.*, Vol. 38, No. 11, pp. 1852–1858, Nov. 1990. doi:10.1109/8.102750

[48] J. M. Jin and J. L. Volakis, "A biconjugate gradient FFT solution for scattering by planar plates," *Electromagnetics*, Vol. 12, No. 1, pp. 105–119, 1992.

[49] S. Washisu, I. Fukai and M. Suzuki, "Extension of finite-element method for unbounded field problems," *Electron. Lett.*, Vol. 15, pp. 772–774, 1979.

[50] T. Orikasa, S. Washisu, T. Honma and I. Fukai, "Finite element method for unbounded field problems and application to two-dimensional taper," *Int. J. Numer. Methods Eng.*, Vol. 19, pp. 157–168, 1983. doi:10.1002/nme.1620190202

[51] S. P. Marin, "Computing scattering amplitudes for arbitrary cylinders under incident plane waves," *IEEE Trans. Antennas Propag.*, Vol. 30, No. 6, pp. 1045–1049, Nov. 1982. doi:1109/TAP.1982.1142939

[52] J. M. Jin and V. V. Liepa, "Application of hybrid finite element method to electromagnetic scattering from coated cylinders," *IEEE Trans. Antennas Propag.*, Vol. 36, No. 1, pp. 50–54, Jan. 1988. doi:10.1109/8.1074

[53] J. D. Angelo, M. J. Povinelli and M. A. Palmo, "Hybrid finite element/boundary element analysis of a strip notch array," *IEEE Antennas Propag. Int. Symp. Dig.*, Vol. 3, pp. 1126–1129, June 1988.

[54] Z. Gong and A. W. Glisson, "A hybrid equation approach for the simulation of electromagnetic scattering problems involving two-dimensional inhomogeneous dielectric cylinders," *IEEE Antennas Propag. Int. Symp. Dig.*, Vol. 38, No. 1, pp. 60–68, Jan. 1990.

[55] K. L. Wu, G. Y. Delisle, D. G. Fang and M. Lecours, *Coupled Finite Element and Boundary Element Methods in Electromagnetic Scattering*. Elsevier, New York, 1990.

[56] X. Yuan, D. R. Lynch and J. W. Strohbehn, "Coupling of finite element and moment methods for electromagnetic scattering from inhomogeneous objects," *IEEE Trans. Antennas Propag.*, Vol. 38, No. 3, pp. 386–393, Mar. 1990. doi:10.1109/8.52246

[57] J. M. Jin and J. L. Volakis, "TE scattering by an inhomogeneously filled aperture in a thick conducting plane," *IEEE Trans. Antennas Propag.*, Vol. 38, No. 8, pp. 1280–1286, Aug. 1990. doi:10.1109/8.56967

[58] —— "TM scattering by an inhomogeneously filled aperture in a thick conducting plane," *IEE Proc. H*, Vol. 137, No. 3, pp. 153–159, June 1990.

[59] J. D. Collins, J. Jin and J. L. Volakis, "A combined finite element-boundary element formulation for solution of two-dimensional problems via CGFFT," *Electromagnetics*, Vol. 10, No. 4, pp. 423–437, Oct. 1990.

[60] J. M. Jin, J. L. Volakis and J. D. Collins, "A finite element-boundary integral method for scattering by two- and three-dimensional structures," *IEEE Antennas Propag. Mag.*, Vol. 33, No. 3, pp. 22–32, June 1991. doi:10.1109/74.88218

[61] M. Costabel, *Symmetric Methods for the Coupling of Finite Elements and Boundary Elements*. Berlin: Springer-Verlag, 1987, pp. 411–420.

[62] D. J. Hoppe, L. W. Epp and J. F. Lee, "A hybrid symmetric FEM/MOM formulation applied to scattering by inhomogeneous bodies of revolution," *IEEE Trans. Antennas and Propagat*, Vol. 42, pp. 798–805, June 1994. http://dx.doi.org/10.1109/8.301698

[63] K. M. Mitzner, "Numerical solution of the exterior scattering problem at eigen-frequencies of the interior problem," in *URSI Radio Sci. Mtg. Dig.*, Sept. 1968, p. 75.

[64] R. Mittra and C. A. Klein, *Stability and Convergence of Moment Method Solutions* in Electromagnetics, Mittra, Ed. Berlin, Springer-Verlag, New York, 1975.

[65] J. R. Mautz and R. F. Harrington, "H-field, E-field, and combined-field solutions for conducting body of revolution," *Arch. Elek. Übertragung*, Vol. 32, pp. 157–164, April 1978.

[66] —— "A combined-source formulation for radiation and scattering from a perfectly conducting body," *IEEE Trans. Antennas Propag.*, Vol. 27, No. 4, pp. 445–454, July 1979. doi:10.1109/TAP.1979.1142115

[67] A. D. Yaghjian, "Augmented electric- and magnetic-field equations," *Radio Sci.*, Vol. 16, pp. 987–1001, Nov. 1981.

[68] P. L. Huddleston, L. N. Medgyesi-Mitschang and J. M. Putnam, "Combined field integral formulation for scattering by dielectrically coated conducting bodies," *IEEE Trans. Antennas Propag.*, Vol. 34, No. 4, pp. 510–520, April 1986. doi:10.1109/TAP.1986.1143846

[69] D. R. Wilton and J. E. Wheeler III, *"Comparison of Convergence rates of the conjugate gradient method applied to various integral equation formulations"*, Progress in Electromagnetics Research, PIER 05, pp. 131–158, 1991.

[70] W. D. Murphy, V. Rokhlin and M. S. Vassiliou, "Solving electromagnetic scattering problems at resonant frequencies," *J. Appl. Phys.*, Vol. 67, pp. 6061–6065, May 1990. doi:10.1063/1.345217

[71] J. D. Collins, J. Jin and J. L. Volakis, "Eliminating interior resonances in finite element-boundary integral methods for scattering," *IEEE Trans. Antennas Propag.*, Vol. 40, No. 12, pp. 1583–1585, Dec. 1992. doi:10.1109/8.204753

[72] S. C. Lee, M. N. Vouvakis K. Zhao and J. F. Lee, "Analysing microwave devices using a symmetric coupling of finite and boundary elements," *Int. J. Numer. Methods Eng.*, Vol. 64, No. 4, pp. 528–546, June 2005. http://dx.doi.org/10.1002/nme.1383

[73] T. B. A Senior and J. L. Volakis, *Approximate Boundary Conditions in Electromagnetics.* IEE Press, London, 1995.

[74] E. Topsakal, "A new circuit model for the analysis of frequency selective surfaces and volumes," Digest, Vol. 2, pp. 2183–2186, *IEEE Antennas and Propagation Society International Symposium*, Monterey, CA, June 2004.

[75] F. Kikuchi, "Mixed and penalty formulations for finite element analysis of an eigenvalue problem in electromagnetism," *Comp. Methods Appl. Mech. Eng.*, Vol. 64, pp. 509–521, 1987. doi:10.1016/0045-7825(87)90053-3

[76] W. H. Press, S. A. Teukolsky, W. T. Vetterling and B. P. Flannery, *Numerical Recipes in C: The Art of Scientific Computing*, 2nd edition. Cambridge University Press, 1992.

[77] M. G. Duffy, "Quadrature over a pyramid or cube of integrands with a singularity at a vertex," *SIAM J. Numer. Anal.*, Vol. 19, No. 6, pp. 1260–1262, Dec. 1982. doi:10.1137/0719090

[78] K. Sertel and J. L. Volakis, "Method of moments solution of volume integral equations using parametric geometry modeling," Radio Science, Vol. 37, No. 1, pp. 1–7, 2002.

[79] H. H. Syed and J.L. Volakis, "2D hybrid FE–BI formulation for anisotropic structures containing pec, impedance, resistive and capacitive surfaces," *Technical Report*, University of Michigan, Radiation Lab., EECS Dept., April 2001.

[80] A. F. Peterson, "Analysis of heterogeneous electromagnetic scatterers: Research progress of the past decade," *Proc. IEEE*, Vol. 79, No. 10, pp. 1431–1441, Oct. 1991. doi:10.1109/5.104218

[81] J. M. Jin and J. L. Volakis, "A finite element-boundary integral formulation for scattering by three-dimensional cavity-backed apertures," *IEEE Trans. Antennas Propag.*, Vol. 39, No. 9, pp. 97–104, Jan. 1991. doi:10.1109/8.64442

[82] —— "Electromagnetic scattering by and transmission through a three-dimensional slot in a thick conducting plane," *IEEE Trans. Antennas Propag.*, Vol. 39, No. 4, pp. 543–550, April 1991. doi:10.1109/8.81469

[83] —— "A hybrid finite element method for scattering and radiation by microstrip patch antennas and arrays residing in a cavity," *IEEE Trans. Antennas Propag.*, Vol. 39, No. 11, pp. 1598–1604, Nov. 1991. doi:10.1109/8.102775

[84] J. L. Volakis, A. Chatterjee and J. Gong, "A class of hybrid finite element methods for electromagnetics: A review," *J. Electromagn. Wave Appl.*, Vol. 8, No. 9/10, pp. 1095–1124, Sept. 1994.

[85] J. L. Volakis, A. Chatterjee and L. C. Kempel, "A review of the finite element method for three dimensional scattering," *J. Opt. Soc. Am. A*, pp. 1422–1433, April 1994.

[86] J. L. Volakis, A. Chatterjee, J. Gong, L. C. Kempel and D. Ross, "Progress on the application of the finite element method to 3D electromagnetic scattering and radiation," *COMPEL—Int. J. Comput. Math. Electr. Electron. Eng.*, Vol. 13, pp. 359–364, May 1994.

[87] J. L. Volakis and A. Chatterjee, "A selective review of the finite element-ABC and the finite element-boundary integral methods for electromagnetic scattering," *Ann. Telecommun.*, Vol. 50, No. 5–6, pp. 499–509, May–June 1995.

[88] J.L. Volakis, J. Gong and T. Ozdemir, *FEM Applications to Conformal Antennas*. Ch. 13 in *Finite Element Software for Microwave Engineering*, Itoh, Silvester, and Pelosi, Ed., pp. 313–345. New York: Wiley, 1996.

[89] R. Coifman, V. Rokhlin and S. Wandzura, "The fast multipole method for the wave equation: A pedestrian prescription," *IEEE Antennas Propag. Mag.*, Vol. 35, No. 3, pp. 7–12, June 1993. doi:10.1109/74.250128

[90] R. L. Wagner and W. C. Chew, "A ray-propagation fast multipole algorithm," *Microw. Opt. Tech. Lett.*, Vol. 7, No. 10, pp. 435–438, July 1994.

[91] S. S. Bindiganavale and J. L. Volakis, "Comparison of three FMM techniques for solving hybrid FE-BI systems," *IEEE Antenna Propag. Mag.*, Vol. 39, No. 4, pp. 47–60, Aug. 1997. doi:10.1109/74.632995

[92] K. Sertel and J. L. Volakis, "Multilevel fast multipole method solution of volume integral equations using parametric geometry modeling," *Antennas and Propagation Society International Symposium, 2001. IEEE* Volume 3, 8–13, July 2001 Page(s):172–175, vol. 3.

[93] R. W. Kindt, K. Sertel, E. Topsakal and J. L. Volakis, "Array decomposition method for the accurate analysis of finite arrays," *IEEE Trans. Antennas Propag.*, Vol. 51, No. 6, pp. 1364–1372, June 2003. doi:10.1109/TAP.2003.811496

[94] Y. Saad, *Iterative Methods for Sparse Linear Systems*. Boston: PWS, 1996.

[95] K. Sertel and J. L. Volakis, "Incomplete LU preconditioner for FMM implementation," *Microw. Opt. Tech. Lett.*, Vol. 26, No. 7, pp. 265–267, 2000.

[96] W. C. Chew, J.-M. Jin, E. Michielssen and J. Song, Eds., *Fast and Efficient Algorithms in Computational Electromagnetics*. Norwood, MA: Artech House, 2001.

[97] S. D. Gedney, A. Zhu and C. C. Lu, "Study of mixed-order basis functions for locally corrected nystrom method," *IEEE Trans. Antennas Propag.*, Vol. 52, No. 11, pp. 2996–3004, Nov. 2004. doi:10.1109/TAP.2004.835122

[98] L. C. Kempel, J. L. Volakis and R. Sliva, "Radiation by cavity-backed antennas on a circular cylinder," *IEE Proc. Microwaves, Antennas and Propagat. H*, Vol. 142, No. 3, pp. 233–239, June 1995.

[99]   T. Özdemir and J. L. Volakis, "Triangular prisms for edge-based vector finite element analysis of conformal antennas," *IEEE Trans. Antennas Propag.*, Vol. 45, No. 5, pp. 788–797, May 1997. doi:10.1109/8.575623

[100] K. Sertel and J. L. Volakis, "Fast Multipole Method (FMM) Solutions of Finite Element-Boundary Integral (FE-BI) Implementations for Antenna Modeling Involving Curved Geometries," *1998 IEEE AP-S International Symposium and URSI Radio Science Meeting*, Atlanta, USA, vol. 1, pp. 232–235, June 1998.

[101] R. W. Kindt and J. L. Volakis, "The array decomposition-fast multipole method for finite array analysis," *Radio Sci.*, Vol. 39, No. 2, April 2004.

[102] T. F. Eibert and J. L. Volakis, "Fast spectral domain algorithms for rapid solution of integral equations," *Electron. Lett.*, Vol. 34, No. 13, pp. 1297–1299, June 1998. doi:10.1049/el:19980934

[103] J. L. Volakis, K. Sertel, E. Jorgensen and R. W. Kindt, "Hybrid finite element and volume integral methods for scattering using parametric geometry," *CMES*, Vol. 5, No. 5, pp. 463–476, 2004.

[104] S. S. Bindiganavale and J. L. Volakis, "Comparison of fast integral mesh truncation schemes for hybrid FE-BI systems," *Electron. Lett.*, Vol. 33, No. 11, pp. 924–925, May 1997. doi:10.1049/el:19970666

[105] B. C. Usner, K. Sertel, M. Carr and J. L. Volakis, "Generalized volume-surface integral equation for modeling inhomogeneities within high contrast composite structures," *IEEE Trans. Antennas Propag.*, Vol. 54, No. 1, pp. 68–75, Jan. 2006. doi:10.1109/TAP.2005.861579

[106] E. Bleszynski, M. Bleszynski and T. Jaroszewicz, "An efficient integral equation based solution method for simulation of electromagnetic fields in inhomogeneous dielectric (biological) media," in *ACES Conf.*, Monterey, CA, Mar. 2000.

[107] G. Kiziltas, D. Psychoudakis, J. L. Volakis and N. Kikuchi, "Topology design optimization of dielectric substrates for bandwidth improvement of a patch antenna," *IEEE Trans. Antennas Propag.*, Vol. 51, No. 10, Part 1, pp. 2732–2743, Oct. 2003. doi:10.1109/TAP.2003.817539

[108] A. J. Poggio and E. K. Miller, *Integral Equation solution of Three Dimensional Scattering Problems*. Elmsford, NY: Permagon, 1973, chapter 4.

[109] Y. Chang and R. F. Harrington, "A surface formulation for characteristic modes of material bodies," *IEEE Trans. Antennas Propag.*, Vol. 25, No. 6, pp. 789–795, Nov. 1977. dx.doi:10.1109/TAP.1977.1141685

[110] T. K. Wu and L. L. Tsai, "Scattering from arbitrarily-shaped lossy dielectric bodies of revolution," *Radio Sci.*, Vol. 12, No. 5, pp. 709–718, Sep-Oct 1977. doi:10.1109/22.60001

[111]   T. K. Sarkar, S. M. Rao and A. R. Djordjević, "Electromagnetic scattering and radiation from finite microstrip structures," *IEEE Trans. Microw. Theory Tech.*, Vol. 38, No. 11, pp. 1568–1575, Nov. 1990.

[112]   S. M. Rao, C. C. Cha, R. L. Cravey and D. L. Wilkes, "Electromagnetic scattering from arbitrary shaped conducting bodies coated with lossy materials of arbitrary thickness," *IEEE Trans. Antennas Propag.*, Vol. 39, No. 5, pp. 627–631, May 1991. doi:10.1109/8.81490

[113]   A. A. Kishk, A. W. Glisson and P. M. Goggans, "Scattering from conductors coated with materials of arbitrary thickness," *IEEE Trans. Antennas Propag.*, Vol. 40, No. 1, pp. 108–112, Jan. 1992. doi:10.1109/8.123366

[114]   B. M. Kolundžija, "Electromagnetic modeling of composite metallic and dielectric structures," *IEEE Trans. Microw. Theory Tech.*, Vol. 47, No. 7, pp. 1021–1032, July 1999. doi:10.1109/22.775434

[115]   L. N. Medgyesi-Mitschang, J. M. Putnam and M. B. Gedera, "Generalized method of moments for three-dimensional penetrable scatterers," *J. Opt. Soc. Am. A*, Vol. 11, No. 4, pp. 1383–1398, April 1994.

[116]   M. Carr, E. Topsakal and J. L. Volakis, "A procedure for modeling material junctions in 3-D surface integral equation approaches," *IEEE Trans. Antennas Propag.*, Vol. 52, No. 5, pp. 1374–1378, May 2004. doi:10.1109/TAP.2004.827247

[117]   J. Shin, A. W. Glisson and A. A. Kishk, "Generalization of surface junction modeling for composite objects in an SIE/MOM formulation using a systematic approach," *ACES J.*, Vol. 20, No. 1, pp. 1–12, Mar. 2005.

[118]   J. H. Richmond, "TE-wave scattering by a dielectric cylinder of arbitrary cross-section shape," *IEEE Trans. Antennas Propag.*, Vol. 14, No. 4, pp. 460–464, July 1966. doi:10.1109/TAP.1966.1138730

[119]   D. E. Livesay and K. M. Chen, "Electromagnetic fields inside arbitrarily shaped biological bodies," *IEEE Trans. Microw. Theory Tech.*, Vol. 22, No. 12, pp. 1273–1280, Dec. 1974. doi:10.1109/TMTT.1974.1128475

[120]   D. H. Schaubert, D. R. Wilton and A. W. Glisson, "A tetrahedral modeling method for electromagnetic scattering by arbitrary shaped inhomogeneous dielectric bodies," *IEEE Trans. Antennas Propag.*, Vol. 32, No. 1, pp. 77–85, Jan. 1984. doi:10.1109/TAP.1984.1143193

[121]   W. C. Chew and C. C. Lu, "The use of hyugen's equivalence principle for solving the volume integral equation of scattering," *IEEE Trans. Antennas Propag.*, Vol. 41, No. 7, pp. 897–904, July 1993. doi:10.1109/8.237620

[122]   W. C. Chew, *Waves and Fields in Inhomogeneous Media*. IEEE Press, New York, 1995.

[123] J. L. Volakis, "Alternative field representation and integral equations for modeling inhomogeneous dielectrics," *IEEE Trans. Microw. Theory Tech.*, Vol. 40, No. 3, pp. 604–608, Mar. 1992. doi:10.1109/22.121745

[124] C. T. Tai, "A note on the integral equation for the scattering of a plane wave by an electromagnetically permeable body," *Electromagnetics*, Vol. 5, pp. 79–88, 1985.

[125] —— "Direct integration of field equations," *Prog. Electromagn. Res.*, Vol. 28, pp. 339–359, 2000. doi:10.2528/PIER99101401

[126] M. I. Sancer, K. Sertel, J.L. Volakis, and P. Van Alstine, "On volume integral equations," *IEEE Trans. Antennas Propag.*, Vol. 54, No. 5, pp. 1488–1495, May 2006. doi:10.1109/TAP.2006.874316

[127] K. Sertel, M. Sancer and J. L. Volakis, "Volume integral equations for permeable structures," *Antennas and Propagation Society International Symposium, 2003. IEEE* Vol. 3, 22–27 June 2003 Page(s): 2–5 vol. 3.

[128] M. Carr, "Scattering from helicopter platforms: Generalized fast methods, preconditioning, and hybrid techniques," Ph.D. dissertation, University of Michigan, Ann Arbor, 2002.

[129] Z. Q. Zhang and Q. H. Liu, "Three-dimensional nonlinear image reconstruction for microwave biomedical imaging," *IEEE Trans. Biomed. Eng.*, Vol. 51, No. 3, pp. 544–548, Mar. 2004. doi:10.1109/TBME.2003.821052

[130] C. C. Lu and W. C. Chew, "A coupled surface-volume integral equation approach for the calculation of electromagnetic scattering from composite metallic and material targets," *IEEE Trans. Antennas Propag.*, Vol. 48, No. 12, pp. 1866–1868, Dec. 2000. doi:10.1109/8.901277

[131] R. Janaswamy, "An accurate moment model for the tapered slot antenna," *IEEE Trans. Antennas Propag.*, Vol. 37, No. 12, pp. 1523–1528, Dec. 1989. doi:10.1109/8.45093

[132] B. E. Fischer, "Electromagnetic design optimization: Application to a patch antenna reflection loss on a textured material substrate," Ph.D. dissertation, University of Michigan, Ann Arbor, 2005.

[133] B. A. Munk, *Frequency Selective Surfaces, Theory and Design*. New York: Wiley, 2000.

[134] —— *Finite Antenna Arrays and FSS*. New York: Wiley, 2003.

[135] R. Mittra, C. H. Chan and T. Cwik, "Techniques for analyzing frequency selective surfaces—a review," *Proc. IEEE*, Vol. 76, No. 12, pp. 1593–1615, Dec. 1988. doi:10.1109/5.16352

[136] S. A. Tretyakov, "Meta-materials with wideband negative permittivity and permeability," *Microw. Opt. Technol. Lett.*, Vol. 31, No. 3, pp. 163–165, Sep. 2001. doi:10.1002/mop.1387

[137] C. L. Holloway, E. F. Kuester, J. B. Jarvis and P. Kabos, "A double negative (DNG) composite medium composed of magnetodielectric spherical particles embedded in a matrix," *IEEE Trans. Antennas Propag.*, Vol. 51, No. 10, Part 1, pp. 2596–2603, Oct. 2003. doi:10.1109/TAP.2003.817563

[138] A. Figotin and I. Vitebskiy, "Electromagnetic unidirectionality in magnetic photonic crystals," *Phys. Rev. B*, Vol. 67, p. 165210, 2003. doi:10.1103/PhysRevB.67.165210

[139] S. D. Gedney and R. Mittra, "Analysis of the electromagnetic scattering by thick gratings using a combined FEM/MM solution," *IEEE Trans. Antennas Propag.*, Vol. 39, No. 11, pp. 1605–1614, Nov. 1991.

[140] J. M. Jin and J. L. Volakis, "Scattering and radiation analysis of three-dimensional cavity arrays via a hybrid finite-element method," *IEEE Trans. Antennas Propag.*, Vol. 41, No. 11, pp. 1580–1586, Nov. 1993. doi:10.1109/8.267360

[141] E. W. Lucas and T. P. Fontana, "A 3D hybrid finite element/boundary element method for the unified radiation and scattering analysis of general infinite periodic arrays," *IEEE Trans. Antennas Propag.*, Vol. 43, No. 2, pp. 145–153, Feb. 1995. doi:10.1109/8.366376

[142] T. F. Eibert, J. L. Volakis, D. R. Wilton, and D. R. Jackson, "Hybrid FE/BI modeling of 3D doubly periodic structures utilizing triangular prismatic elements and an MPIE formulation accelerated by the Ewald transformation," *IEEE Trans. Antennas Propag.*, Vol. 47, No. 5, pp. 843–850, May 1999. doi:10.1109/8.774139

[143] T. F. Eibert and J. L. Volakis, "Fast spectral domain algorithm for hybrid finite elements/ boundary integral modeling of doubly periodic structures," *IEE Proc. Microw. Antennas Propag.*, Vol. 147, No. 5, pp. 329–334, Oct. 2000. doi:10.1049/ip-map:20000706

[144] B. C. Usner, K. Sertel and J. L. Volakis, "Doubly periodic volume-surface integral equation formulation for modeling metamaterials," *IEE Proc. Microw. Antennas Propag.*, to appear in Dec. 2006.

[145] P. P. Ewald, "Die berechnung optischer und elektrostatischen gitterpotentiale," *Ann. Phys.*, Vol. 64, pp. 253–258, 1921.

[146] K. E. Jordan, G. R. Richter and P. Sheng, "An efficient numerical evaluation of the Green's function for the helmholtz operator on periodic structures," *J. Comput. Phys.*, Vol. 63, pp. 222–235, 1986. doi:10.1016/0021-9991(86)90093-8

[147] J. L. Volakis, T. F. Eibert and K. Sertel, "Fast integral methods for conformal antenna and array modeling in conjunction with hybrid finite element formulations," *Radio Sci.*, Vol. 35, No. 2, pp. 537–546, Mar. 2000. doi:10.1029/1999RS900050

[148] T. F. Eibert, Y. E. Erdemli and J. L. Volakis, "Hybrid finite element-fast spectral domain multilayer boundary integral modeling of doubly periodic structures," *IEEE Trans. Antennas Propag.*, Vol. 51, No. 9, pp. 2517–2520, Sept. 2003. doi:10.1109/TAP.2003.816386

[149] H. Aroudaki, V. Hansen, H.-P. Gemund and E. Kreysa, "Analysis of low-pass filters consisting of multiple stacked fss's of different periodicities with applications in the submillimeter radioastronomy," *IEEE Trans. Antennas Propag.*, Vol. 43, No. 12, pp. 1486–1491, Dec. 1995. doi:10.1109/8.475943

[150] H-Y. D. Yang, R. Diaz and N. G. Alexopoulos, "Reflection and transmission of waves from multilayer structures with planar-implanted periodic material blocks," *J. Opt. Soc. Am. B*, Vol. 14, No. 10, pp. 2513–2521, Oct. 1997.

[151] G. Milton, *The Theory of Composites*. Cambridge University Press, 2002.

[152] L. Lewin, "The electrical constants of a material loaded with spherical particles," *Proc. Inst. Electr. Eng.*, Vol. 94, pp. 65–68, 1947.

[153] V. G. Veselago, "The electrodynamics of substances with simultaneously negative values of ε and μ," *Usp. Fiz. Nauk.*, Vol. 92, pp. 517–526, 1967.

[154] L. Jylha, I. Kolmakov, S. Maslovski and S. Tretyakov, "Modeling of isotropic backward-wave materials composed of resonant spheres," eprint arXiv: cond-mat/0507324v1, July 2005.

[155] G. V. Eleftheriades and K. G. Balmain, *Negative Refraction Metamaterials: Fundamental Principles and Applications*. IEEE Press: Hoboken, N.J: Wiley-Interscience, 2005.

[156] P. Papalambros and D. Wilde, *Principles of Optimal Design: Modeling and Computation*, 2nd edition. New York: Cambridge University Press, 2000.

[157] S. Rao, *Engineering Optimization: Theory and Practice*, 3rd edition. New York: Wiley, 1996.

[158] T. Nakata and N. Takahashi, "New design method of permanent magnets by using the finite element method," *IEEE Trans. Magn.*, Vol. 19, No. 6, pp. 2494–2497, 1983. doi:10.1109/TMAG.1983.1062870

[159] G. Kiziltas, N. Kikuchi, J. L. Volakis and J. Halloran, "Dielectric material optimization for electromagnetic applications using SIMP," *Arch. Comput. Methods Eng.*, Vol. 11, No. 4, pp. 355–388, 2004.

[160] K. Schittkowski, "On the convergence of a sequential quadratic programming method with an augmented Lagrangian line search function," *Math. Operationsforsch. Statist, Ser. Optim*, Vol. 14, pp. 197–216, 1983.

[161] M. Powell, "A fast algorithm for nonlinearly constrained optimization calculations," in *Numerical Analysis*, (Lecture Notes in Mathematics), Vol. 630. Berlin: Springer, 1978.

[162] J. Holland, *Adaptation in Natural and Artificial Systems*. Ann Arbor: University of Michigan Press, 1975.

[163] J. M. Johnson and Y. Rahmat-Samii, "Genetic algorithm optimization and its application to antenna design," *IEEE Antennas Propag. Int. Symp.*, Seattle, WA, June 1994.

[164] E. Aarts and J. Korst, *Simulated Annealing and Boltzman Machines: A Stochastic Approach to Combinatorial Optimization and Neural Computing*. New York: Wiley, 1988.

[165] H. J. Delgado and M. H. Thursby, "A novel neural network combined with FDTD for the synthesis of a printed dipole antenna," *IEEE Trans. Antennas Propag.*, Vol. 53, No. 7, pp. 2231–2236, July 2005. doi:10.1109/TAP.2005.850706

[166] D. Srinivasan, S. Ratnajeevan and H. Hoole, "Fuzzy multiobject optimization for the starting design of a magnetic circuit," *IEEE Trans. Magn.*, Vol. 32, No. 3, pp. 1230–1233, May 1996. dx.doi:10.1109/20.497466

[167] J. Robinson and Y. Rahmat-Samii, "Particle swarm optimization in electromagnetics," *IEEE Trans. Antennas Propag.*, Vol. 52, No. 2, pp. 397–407, Feb. 2004. doi:10.1109/TAP.2004.823969

[168] D. Goldberg, *Genetic Algorithms in Search, Optimization, and Machine Learning*. Reading, MA: Addison-Wesley, 1989.

[169] S. Kirkpatrick, C. Gelatt Jr. and M. Vecchi, "Optimization by simulated annealing," *Science*, Vol. 220, No. 4598, pp. 671–680, 1983.

[170] A. Jones and G. Forbes, "An adaptive simulated annealing algorithm for global optimization over continuous variables," *J. Glob. Opt.*, Vol. 6, No. 1, pp. 1–37, Jan. 1995. doi:10.1007/BF01106604

[171] L. Lebensztajn, C. A. R. Marretto, M. C. Costa and J. L. Coulomb, "Kriging: a useful tool for electromagnetic devices optimization," *IEEE Trans. Magn.*, Vol. 40, No. 2, pp. 1196–1199, July 2004. http://dx.doi.org/10.1109/TMAG.2004.824542

[172] E. S. Siah, M. Sasena, J. L. Volakis, P. Y. Papalambros and R. W. Wiese, "Fast parameter optimization of large-scale electromagnetic objects using DIRECT with kriging meta-modeling," *IEEE Trans. Microw. Theory Tech.*, Vol. 52, No. 1, pp. 276–285, Jan. 2004. doi:10.1109/TMTT.2003.820891

[173] G. I. N. Rozvany, M. Zhou and T. Birker, "Generalized shape optimization without homogenization," *Struct. Opt.*, Vol. 4, No. 3–4, pp. 250–252, 1992. doi:10.1007/BF01742754

[174] O. Sigmund, "Design of multi-physics actuators using topology optimization—Part I: One material structures," *Comput. Methods Appl. Mech. Eng.*, Vol. 190, pp. 6577–6604, 2001. doi:10.1016/S0045-7825(01)00251-1

[175] J. M. Johnson and Y. Rahmat-Samii, "Genetic algorithms and method of moments (ga/mom) for the design of integrated antennas," *IEEE Trans. Antennas Propag.*, Vol. 47, No. 10, pp. 1606–1614, Oct 1999. doi:10.1109/8.805906

[176] D. N. Dyck, D. A. Lowther and E. M. Freeman, "A method of computing the sensitivity of electromagnetic quantities to changes in materials and sources," *IEEE Trans. Magn.*, Vol. 30, No. 5, pp. 3415–3418, 1994. http://dx.doi.org/10.1109/20.312672

[177] H. L. Thomas, G. N. Vanderplaats and Y. K. Shyy, "A study of move-limit adjustment strategies in the approximation concepts approach to structural analysis," in *4th AIAA/USAF/NASA/OAI Symp. on Multidisciplinary Analysis and Optimization*, Cleveland, Ohio, 1992.

[178] Z. Li, Y. E. Erdemli, J. L. Volakis and P. Y. Papalambros, "Design optimization of conformal antennas by integrating stochastic algorithms with the hybrid finite-element method," *IEEE Trans. Antennas Propag.*, Vol. 50, No. 5, pp. 676–684, May 2002. doi:10.1109/TAP.2002.1011234

[179] S. Koulouridis, D. Psychoudakis and J. L. Volakis, "Multi-objective optimal antenna design based on volumetric material optimization," *IEEE Trans. Antennas Propag.*, (special issue on Synthesis and Optimization Techniques in Electromagnetics and Antenna System Design), to appear in 2007.

[180] H. Choo, R. L. Rogers and H. Ling, "Design of electrically small wire antennas using a Pareto genetic algorithm," *IEEE Trans. Antennas. Propag.*, Vol. 53, No. 5, pp. 1038–1046, March 2005. doi:10.1109/TAP.2004.842404

Biography

John L. Volakis is a native of Greece and immigrated to the United States when 18 years old. He obtained his Ph.D. from the Ohio State University in 1982, and after a short term (1982–84) at Rockwell International (now Boeing Phantom Works), he was appointed Assistant Professor at the University of Michigan (Ann Arbor, Michigan), becoming a Professor in 1994. Since January 2003 he is the Roy and Lois Chope Chair Professor of Engineering at the Ohio State University, Columbus, Ohio and also serves as the Director of the ElectroScience Laboratory. Prof. Volakis has published 230 refereed journal articles, nearly 350 conference papers and 10 book chapters. He co-authored 2 other books: *Approximate Boundary Conditions in Electromagnetics* (Institution of Electrical Engineers, London, 1995), *Finite Element Method for Electromagnetics* (IEEE Press, New York, 1998). In recent years, he has been listed by ISI as one among the most referenced authors in computer science/engineering. He graduated/mentored nearly 50 Ph.D. students/post-docs, and co-authored with them 5 best paper awards at conferences. Dr. Volakis is a Fellow of the IEEE, and served as the 2004 President of the IEEE Antennas and Propagation Society. He was Associate Editor for several journals and was twice the general chair/co-chair of the IEEE Antennas and Propagation conference.

Kubilay Sertel was born in 1973 in Tekirdag, Turkey. He received his B.S. and M.S. in Electrical and Electronics Engineering from the Middle East Technical University (Ankara, Turkey) in 1995 and Bilkent University (Ankara, Turkey) in 1997, respectively. He received his Ph.D. in 2003 from the University of Michigan (Ann Arbor, Michigan). He subsequently joined the ElectroScience Laboratory at the Ohio State University (Columbus, Ohio) as a Senior Research Associate where he is also an Adjunct Assistant Professor in the Electrical and Computer Engineering Department since 2005. He has co-authored 20 refereed journal papers and over 50 conference papers. His expertise is on fast and efficient numerical solutions of surface and volume integral equations as well as hybrid finite element methods. Computer codes developed by Dr. Sertel are being used in academic research and have been adapted to industry applications.

Brian C. Usner is a native of New Orleans, Louisiana. He obtained his B.S. in electrical engineering from Tulane University (New Orleans, LA) in 2001, his M.S. in electrical engineering from the University of Michigan (Ann Arbor, MI) in 2002, and his Ph.D. from the Ohio State University (Columbus, OH) in 2006. Dr. Usner was a Departmental Fellow at the University of Michigan, a University Fellow at the Ohio State University, and a NASA Graduate Student Research Program (GSRP) Fellow at the NASA Langley Research Center in Langley, VA. Dr. Usner was awarded a best paper award at the Applied Computational Electromagnetic Society (ACES) conference in 2006. Professionally, Dr. Usner has worked as a Scientist at BBN Technologies (Cambridge, MA) and is currently a Research and Development Engineer at Ansoft Corporation (Pittsburgh, PA).

Printed in the United States
by Baker & Taylor Publisher Services